T0318813

Neurofeedback

Series Editor
Bruno Salgues

Neurofeedback

Tools, Methods and Applications

Pascale Vincent

First published 2018 in Great Britain and the United States by ISTE Press Ltd and Elsevier Ltd

ISTE Press Ltd
27-37 St George's Road
London SW19 4EU
UK

www.iste.co.uk

Elsevier Ltd
The Boulevard, Langford Lane
Kidlington, Oxford, OX5 1GB
UK

www.elsevier.com

Notices

Knowledge and best practice in this field are constantly changing. As new research and experience broaden our understanding, changes in research methods, professional practices, or medical treatment may become necessary.

Practitioners and researchers must always rely on their own experience and knowledge in evaluating and using any information, methods, compounds, or experiments described herein. In using such information or methods they should be mindful of their own safety and the safety of others, including parties for whom they have a professional responsibility.

To the fullest extent of the law, neither the Publisher nor the authors, contributors, or editors, assume any liability for any injury and/or damage to persons or property as a matter of products liability, negligence or otherwise, or from any use or operation of any methods, products, instructions, or ideas contained in the material herein.

For information on all our publications visit our website at http://store.elsevier.com/

British Library Cataloguing-in-Publication Data
A CIP record for this book is available from the British Library
Library of Congress Cataloging in Publication Data
A catalog record for this book is available from the Library of Congress
ISBN 978-1-78548-276-2

Printed and bound in the UK and US

Contents

Foreword

Using the brain's natural architecture

In 1929, when the neuropsychiatrist Hans Berger recorded the first sign of electrical activity in the brain, thereby identifying the existence of brain waves and more specifically those on the alpha and beta frequencies, he could not have imagined the extent of the future discoveries concerning how the brain functions.

We are now far from René Descartes' concept of the "brain machine"!

The work of Hans Berger was taken up by the Englishman Edgar Adrian, which led to the development of the electroencephalogram (EEG) in the 1950s and its widespread use in medical practice.

Between 1970 and 1980, another huge milestone was reached in neurobiology with the discovery of the brain's neuroplasticity, notably through the work of neurologist Michael Merzenich. This was one of the most important discoveries in neuroscience because it demonstrated the nonlinear and dynamic nature of the brain, possibly the most complex organism in the universe, an organism that is in constant reconfiguration.

This is where *neurofeedback* draws our attention. In my clinical research on anxiety and mood disorders, I focus on non-invasive techniques, including neurofeedback. By sending the brain information about the variations in its activities, we ought to be able to help the brain regulate itself.

It has been observed that mice and rats have the same pharmacological profiles and the same brain oscillations as humans. Every psychiatric disorder could therefore be associated with a sort of malfunction of certain rhythms within the brain, and among the many self-organized rhythms that the brain generates, there is certainly a "neural syntax" to discover.

Medical imaging has led researchers to believe that there are a large number of neural networks in the human brain. Some seem synchronized in their activities. Moreover, neural activity in the brain gives rise to transmembrane currents that can be measured in an extracellular environment and processed mathematically.

In view of the above, neurofeedback is headed in the right direction.

Until now, there has been no published work from which to learn about the fundamentals of this new technique: neurofeedback. This book answers essential questions on the topic and invites us to participate in the daily experience of a practitioner: Pascale Vincent. We are pleased to see this technique, which has been in use for more than 30 years in the United States, appear in France to support treatments related to psychotherapy as well as cognitive rehabilitation.

Let us hope that she confirms that it could help patients who are suffering!

Dr Elie HANTOUCHE[1]

---

1 Dr Elie Hantouche is a psychiatrist who has specialized in bipolar and obsessive–compulsive disorders for more than 20 years. He runs the *Centre des troubles anxieux et de l'humeur* (Center for Anxiety and Mood Disorders) in Paris.

A Note to the Reader

The reader may be confused by the vocabulary used in this book. I use the terms "clients" and "patients" interchangeably to designate people who come to see me.

The word "patient" comes from the Latin word *patiens*, which means "one who suffers". Some of the people who come to see me do indeed suffer. Therefore, this term seems appropriate.

The word "client" is more commonly used in English-speaking cultures. Given that people who resort to neurofeedback also pay for their sessions, this term cannot be excluded.

It seems to me that both terms are appropriate for the practice of neurofeedback, all the more so because I am myself a practitioner or clinician depending on the system that I use: a practitioner of *dynamic neurofeedback* and a clinician of neurofeedback/EEG-biofeedback, which is called *classic neurofeedback* in this book.

We are also starting to use the term "actient" in French to refer to "a patient who acts" due to the increase in the number of patients who find information themselves (thanks to the Internet among other resources), and ask practitioners more and more questions, thereby becoming actors in their therapy.

"Clients–patients" or "patients–clients" become actors in their health and seek out more solutions by themselves. This new term of "actients" will most likely become a part of common language within the next few years and corresponds absolutely to people who resort to neurofeedback.

Preface

I live in Séné, on the Gulf of Morbihan in Brittany, France. It is said to be one of the most beautiful bays in the world. The site is also classified as a natural regional park.

Morbihan means "the little sea" in Breton. One of the particularities of this inland sea is the very narrow gully that allows the sea to enter the gulf. This passage between Port-Navalo and Locmariaquer (near the island of La Jument) is where the strongest tidal currents in Europe are recorded. This feature, together with the numerous islands in the gulf, means that the tidal schedules are more delayed than on the coast where there are almost no obstacles to the movements of the sea.

The flow of water is so significant on the turn of the tide that it creates a very powerful current that many boats struggle against with difficulty. The current's power creates what could be called a "disruption" in tidal time.

While high tide is at about the same time, give or take a few minutes, at different points along France's Atlantic coast despite them being several hundreds of kilometers apart, there is a two-hour delay between the entrance to the gulf and the end of the gulf. The towns of Port-Navalo and Vannes, however, are only about ten kilometers apart, as the crow flies.

In the Gironde estuary, in addition to a long delay between the tide times, there is a phenomenon of "clashing tides". When the tide ebbs, it flows in the outgoing direction of the Garonne and the Dordogne in the Bay of Biscay. On the other hand, when the tide comes in, the two flows are opposed, thus disrupting the tidal movements. This leads to a wide variation in the water levels recorded at different moments of the cycle.

At Mont-Saint-Michel, the flatness of the surrounding area creates an extremely large range between high tide and low tide, as no obstacles hold back, stop or divert the sea in its progression. The sea's movements at the time of the rising and ebbing tide do not clash.

All of these phenomena, although quite different, are natural and depend entirely on the specific nature of the topography of each of these places along the French coast. There is nothing abnormal about these very remarkable maritime phenomena.

Tides occur for several reasons:

– the rotation of the Earth which, through centrifugal force, creates a bulge of water in the oceans. This bulge tends to diminish further away from the equator and is no longer formed close to the poles: there is only one tide per day in Australia;

– the attraction of the Moon on the Earth also creates a more or less large bulge in the water depending on the phases of the Moon and its proximity to the Earth;

– the attraction of the Sun on the Earth which also creates a bulge in the water toward the solar star.

All of these phenomena are natural, cyclical and predictable.

Some factors that are independent of the tidal system can act on and influence the range of the tide in a localized and specific way, especially wind, rain and atmospheric pressure.

When the atmospheric pressure decreases, this signals the arrival of a meteorological disturbance that can modify the stable conditions of the rhythm of the tides at the geographical point where it occurs. The arrival of a depression is synonymous with clouds, rain, wind and, if the depression is too strong, a storm.

Sometimes, weather disturbances occur in several areas around the globe simultaneously in a more or less related way. Some years see many extreme weather phenomena.

In the winter of 2010, in Vendée, a tidal wave triggered by the cyclone Xynthia caused floods and extensive damage to homes.

It was a combination of several factors that led to this natural catastrophe: strong winds, blowing in the direction of the rising tide, accompanied by torrential rain and a particularly high tide led to an overflow of normal limits of the highest tides.

That same year, the following events occurred around the world:

– on January 12, in Haiti, a magnitude 7 earthquake;

– on February 27, in Chile, a magnitude 8.8 earthquake and cyclone Xynthia in Europe;

– on April 13, in China, a magnitude 7.1 earthquake;

– on June 15, fatal floods in the south of France;

– on July 26, fatal floods in Pakistan;

– on July 29, in Russia, an exceptional drought, the likes of which had not been recorded for 1,000 years, causing severe fires;

– on September 4, in New Zealand, a magnitude 7 earthquake;

– on October 17, strong floods in southern Russia;

– on October 26, in Indonesia, the Merapi volcano erupted.

The Earth has suffered and continues to suffer climatic disturbances that destabilize ecosystems and the rhythm of life for its inhabitants. These disturbances are sometimes devastating.

Humans never cease predicting these events, preparing for them, improving detection systems for natural disasters, strengthening homes, tirelessly protecting and rebuilding what these catastrophes have destroyed, and healing the wounds that they have left on the environment...and on minds.

Introduction

Humans take an interest in their environment to better prepare for it and respond to its potential imbalances.

The brain, like the ocean, is subject to rhythms and regular variations, and also suffers disturbances related to many interior and exterior factors.

Humans are studying the brain and how it functions, but the brain still harbors many secrets: psychological disorders still frighten us, and justifiably so, because they reflect a destabilization in this mysterious organ with immense and fascinating powers.

Neuroscience focuses on how the brain functions and its interactions with the interior world, the organism, and the exterior world, the environment.

Many tools are currently available to observe, heal, calm or optimize cerebral capacities. I would like to provide more information about one of these tools: neurofeedback.

Increasingly often, we hear talk of neurofeedback on the television or radio, and in the scientific or general press as well as on social networks, we can read more or less well-documented and serious articles on the subject.

The mixed views and contradictory opinions, notably in the medical community, have prompted me to make my own personal contribution to the understanding of its function and its use.

The goal of this book is not to trace the history of neurofeedback, which is accessible on any site dedicated to this method, regardless of the type of neurofeedback on offer.

This book should be taken as a technical and informative approach based on personal empirical experience. It is intended to provide information about different methods of neurofeedback, their major areas of interest, and their effects on health and well-being that can be expected.

The content of this work is a reflection on my perception of neurofeedback and the use that I make of it, and is my personal opinion. It documents what I have learned during my training and years of practice, my personal investigations on the topic and experiments I have conducted with my patients during neurofeedback sessions.

But first…

Part 1

Tools and Methodologies

What is Neurofeedback?

Neurofeedback is a generic term that designates a computerized method of brain observation and training originating from neuroscience. The aim is to record and analyze the activity of neurons in order to optimize the brain's performance by showing it information about its own functioning. The information collected, called *feedback*, can be auditory, visual, or even tactile, depending on the neurofeedback system used.

All neurofeedback systems have similar basic functions, with a few variations, but they have different graphical environments, different presentations, a specific design and their own levels of training and performance.

What is fundamental and common to all neurofeedback systems is the reception of the signal emitted by the brain, its transformation, its analysis by the software and its real-time rendering via the feedback.

1.1. Different tools

This book references two therapeutic neurofeedback systems that I use in my own consultation practice:

– the Cygnet® neurofeedback/EEG biofeedback system based on the Othmer method, EEG Info, a first-generation system (created in the 1980s under the name of the EEG Spectrum, which became EEG Info in 2002) and presented, in this book, under the name of *classic neurofeedback*;

– the NeurOptimal® system based on the Zengar method, a second-generation system that developed out of *classic neurofeedback* (created in 1996 under the name of NeuroCare®, which later became NeurOptimal®). In this book, it is called *dynamic neurofeedback*.

There are several other kinds of tools to observe electrical neural activity designed by companies involved in neuroscience research and also in connection with the medical community. Not all of the systems presented below are intended for neurotherapy, unlike the Cygnet® and NeurOptimal® systems.

Some of the most well-known medical instruments used to observe the brain include:

– EEG: electroencephalogram;

– fMRI: functional magnetic resonance imaging;

– MEG: magnetoencephalography;

– fNIRS: functional near-infrared spectroscopy;

– HEG: hemoencephalography.

And among the instruments that can be used to train the brain, aside from the two tools cited above that I use in my own practice:

– LoRETA: *low-resolution electromagnetic tomography.*

These tools allow brain activity to be "visualized" thanks to blood flow (like magnetic resonance imagery) and also thanks to the modification of the impedance of the skin which signals an electrical activation of the brain in one zone or another (magnetoencephalography). The feedback sent is addressed to the therapist and not the patient, unlike therapeutic neurofeedback.

Concretely, brain imaging will inform the therapist about the areas of blood flow in the brain and provide information about how it operates. It is not intended, *a priori*, to be used by patients directly in order for them to act on their own brains to modify their functioning, although this is entirely possible.

Therapeutic neurofeedback informs the practitioner about the cerebral activity of the patient, but above all, it informs patients about their own brain function, allowing them to eventually try to self-regulate it where appropriate. The process can be referred to as a re-education of the brain.

My approach as a neurofeedback clinician is to seek to understand, decode, know and discover a new and fascinating world. Accompanying my patients toward a greater well-being, helping them to understand themselves and find themselves defines my personal approach to this therapeutic method.

Some research centers, in partnership with companies, are developing and experimenting with therapeutic neurofeedback techniques in order to commercialize them for the general public. This is the case of the INRIA (French Institute for Research in Computer Science and Automation) and the company Mensia Technologies which has already developed neurotherapy tools like the MENSIA KOALA^{TM} that targets ADHD. These research groups are not the only ones – far from it – to focus on neuroscience.

However, the brain–machine interface is not only used in the medical field. It is the subject of research in many domains outside of health, notably in video games in augmented reality, for example, which are currently booming.

1.2. How a session unfolds

The first session lasts between one hour and one and a half hours.

During the initial meeting, the patient fills out a questionnaire (history) about various problems that could concern them. The goal of this questionnaire is to then evaluate the changes that are produced. An evolution assessment can be taken after a few brain training sessions, but the brain requires time to regulate itself: the effects of neurofeedback are very rarely immediate.

After placing the electrodes, a first measurement (baseline) is taken of the brain's electrical activity that makes it possible to assess its state (anxiety, stress, emotionality, etc.) to choose the appropriate session. This also makes it possible to verify that the electrodes are correctly positioned and are receiving a good signal.

After this first analysis, the person chooses the music or the video support that they want for the brain training session, which lasts about half an hour (Figure 1.1).

Then, the patient listens to the music and/or watches a video program *(dynamic neurofeedback)*, or carries out tasks appropriate to the session if it consists of active neurofeedback *(classic neurofeedback)*, while the software records the neural activity in the cerebral area where the electrodes are placed. A disturbance in the sound and image signals an imbalance to the brain. This is the feedback.

During the session, the person perceives these disturbances in sound and image when the software detects too much variation in the electrical signal. The person's brain reacts to these disturbances and attempts to regulate itself. Each modification of the sound and image is perceived as a strong alert signal because the brain does not expect it, especially as it is correlated to its own functioning. This is the information return or feedback.

It is recommended to move as little as possible and relax so that the software does not have to process interference caused by movements or speech, for example.

© Samuel VINCENT

Figure 1.1. *Example of a neurofeedback session. For a color version of this figure, see www.iste.co.uk/vincent/neurofeedback.zip*

The *classic neurofeedback* system has – in addition to feedback – a reward system that signals to the brain when it maintains a stable equilibrium. This reward can be auditory, visual or tactile.

Feedback signals an imbalance and therefore a modification to carry out; the reward indicates an adapted response and a stable and balanced state to maintain, and even to reinforce.

At the end of the session, the patient and the practitioner exchange their impressions: the person can talk about their experience, the mental images or bodily sensations that they may have perceived. The practitioner explains their analysis of different markers on the screen, which can allow the person to understand what happened, better comprehend their own functioning better, and find ways to move forward more efficiently.

Depending on the system, the practitioner may carry out a second measurement of the brain's state (*dynamic neurofeedback*) in order to potentially establish a preliminary assessment of effects that are likely to occur and draw a first draft of the changes that are possible.

Possible does not mean that these changes will necessarily take place!

The following sessions last one hour. To better support this process, the practitioner considers the patient's evolution and experience. The practitioner then adapts the next session accordingly.

As time and the sessions progress, the brain learns and modifies itself, but how the system functions often appears to be mysterious.

Most patients and their families are curious to know how it works and why the brain reacts to the feedback.

A doctor friend told me one day:

"It's a wonderful tool, but I cannot recommend it because I do not understand how it works. If someone asked me, I would not know what to say."

To which one of his colleagues responded:

"From the moment that it works, it takes off! Do we really know how most of the medications that we prescribe work? And yet!"

Here is what, for my part, I understand about how the two neurofeedback systems that I use function.

2

Signal Processing

Most people arriving at my consultation practice for the first time are fairly anxious about the idea of having electrodes placed on their heads. Whether adults or children, the majority of them are convinced that they will transmit electricity to the brain.

Neurofeedback has nothing in common with sismotherapy (also known as electroshock therapy) that some of my patients have experienced – retaining residual effects from that experience, notably in the short-term memory – nor does it bear any relation to transcranial stimulation whose results remain inferior, it seems, to those of neurofeedback, according to patients who have undergone both that technique and also neurofeedback.

These two techniques, sismotherapy and transcranial stimulation, practiced exclusively by hospital personnel, are intrusive methods, unlike neurofeedback. They also have the non-negligible disadvantage of sometimes leaving residual effects, even if they also provide, in some cases, a solution to very complex problems.

I often have to reassure people by explaining to them what neurofeedback is, as well as what it is not. It is not an invasive or intrusive procedure, nor a tool that transmits electricity. In this work, I use the word "electrode" because that is the appropriate name for the sensors that are positioned on the patient's head when recording an EEG.

2.1. Receiving and processing the signal

Electrodes resemble miniature stethoscopes and are positioned on the scalp with a conductive paste. The position and number of electrodes vary from one neurofeedback method to another (two in *dynamic neurofeedback*, four or more in *classic neurofeedback*). They are connected by an amplifier-encoder that is connected to a computer.

The amplifier-encoder receives signals emitted by the brain where the electrodes are positioned in an area of about 4 cm in diameter and a depth of about 3-4 cm. These signals, called brain waves, are amplified and digitized by the amplifier-encoder to reach the neurofeedback software (Figure 2.1).

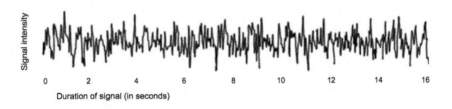

Figure 2.1. *Signal recorded for 16 seconds. For a color version of this figure, see www.iste.co.uk/vincent/neurofeedback.zip*

A Fourier transform allows the software to display the electrical brain activity recorded locally under the electrodes as frequencies (hertz). The Fourier *transform*[1] is used in every system that translates signals into frequencies (Figure 2.2).

If we compare the brain activity to an orchestra, the Fourier transform makes it possible to break down the sounds made by all of the musical instruments and to display each of their scores.

---

1 Jean-Baptiste Fourier: French mathematician and physicist (1768–1830). The Fourier transformation associates an integrable function with a function called a Fourier transform whose independent variable can be interpreted as the frequency. The Fourier transform is expressed as the "infinite sum" of the trigonometric functions of all frequencies.

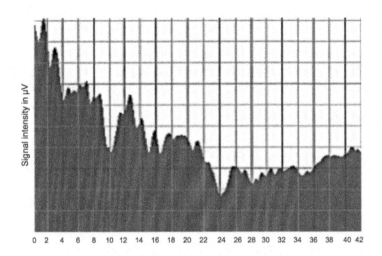

Figure 2.2. *Fourier transformation of the previous signal (in hertz), presented in Figure 2.1*

Regarding the brain, the electrical activity of neurons, expressed in μV (microvolts), is digitized in order to produce a tracing called an electroencephalogram (EEG). The signal sensed by the electrodes is the sum of the electrical pulses emitted by the brain. These pulses are each represented by a sinusoidal function (Figure 2.3).

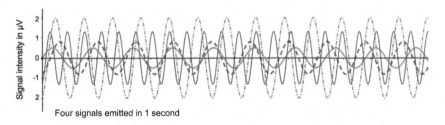

Four signals emitted in 1 second

Figure 2.3. *Breakdown of a signal comprised of four frequencies*

Neurofeedback is not an EEG. It processes information provided by the EEG at points where the electrodes are positioned. This information is amplified in order to be perceptible and exploitable and then broken down into frequencies from 1 to 40 Hz in *classic neurofeedback* or from 1 to 42 Hz and above in *dynamic neurofeedback* (which offers the possibility of also training brain waves higher than 40 Hz called "gamma waves").

2.2. The sampling frequency

The sampling frequency of the signal reception is 256 measurements per second. These measurements, repeated throughout the duration of the neurofeedback session, provide very precise information about the patient's brain activity where the electrodes are positioned. We could call this a "localized" EEG.

The *sampling frequency*[2] described by Shannon is used in very diverse situations and is not limited to EEG measurements.

"Paul, do you remember our vacation by the seaside four years ago?"

"Oh yes, of course!"

"And yet, you were very young..."

"But I remember. It's funny, sometimes it was high, sometimes it was low. One day, we wanted to collect clams and the water was too high, but the week before we had gone together and it was low."

"That's because of the tides. Before going to collect clams, it's a good idea to find out if the sea will be low and what the coefficient is."

"The coefficient?"

"It's an indicator that tells us if the sea will rise very high or be very low."

"How do we find that out?"

"By looking at the tidal schedule."

---

2 Sampling consists of gathering the value of a physical quantity at regular intervals. The sampling frequency is the number of samples per unit of time. If the unit of time is one second, the sampling frequency is expressed in hertz and represents the number of samples measured per second.

"Oh? So the sea is like a train, and we have to look up at what time it arrives and what time it leaves so we don't miss it? But, the train that goes to my grandparents' house leaves every Saturday at the same time."

"That's right. To find out the sea level, you need a tidal schedule. Calculating the tide times is done by measuring the height of the sea at a given point."

"Is it measured with a ruler?"

"Why not? A very large graduated rule, planted vertically in the ground. For each measurement, we observe the water level on the ruler. There are 24 hours in a day."

"Oh, I knew that already!"

"It is high (or low) twice a day. But what we want to know is the state of the sea hour by hour in order to determine a schedule, and not just when it is high and when it is low. There's no use measuring continually: the sea doesn't rise that quickly. By repeating the measurement several times a day for several days in a row, we gradually gather more information about the tides throughout the year."

"But do you have to measure it every day for your whole life?"

"No, of course not! The tides are caused by the phases of the Moon (among other causes). These phases are cyclic. That means that they are repeated from one lunar cycle to another. Look, Paul, here is a tidal schedule. This way, you can go gather clams without being disappointed when we are on vacation in two months!"

"We're going back to the seaside? Awesome! I will find some clams if I can figure out where they hide, but that's a different problem! Thanks Dad!"

The *Shannon sampling theorem*[3], explained simply to little Paul by his father, indicates that in order to get exact measurements of an event without any information being lost, a sufficient number of measurements must be taken.

Shannon demonstrated that to obtain precise information, it is necessary to collect a minimum of twice as many measurements as the maximum frequency of the signal to be measured.

In the brain, the maximum frequency measured is 128 Hz, which is 128 pulses per second. These are infra-low frequencies.

In comparison:

– the main current produced in our homes is a polarized alternating current (+ and −) at a frequency of 50 Hz (in Europe);

– sounds (vibrations of the air) can be perceived by the human ear between 20 and 20,000 Hz;

– FM radio waves (frequency modulation) vary between 87.5 and 108 MHz;

– computer processors today have a cycle time of almost 2 GHz.

CLARIFICATION.– A hertz is a unit of measurement of frequency per second, or in other words, the number of repetitions of a periodic phenomenon over one second, and whose graphical representation has the form of a sinusoidal curve. However, the brain's electrical activity cannot be compared to music waves, for example, because the reference standard is different: in the first case, it consists of pulses measured for one second, and in the second case, it consists of measured vibrations in the air, also for one second (the brain does not produce sound).

---

3 Claude Elwood Shannon: American electrical engineer and mathematician (1916–2001). He is the father of information theory. He demonstrated that, to be as close as possible to the signal without any information being lost, the sampling frequency (F_e) must be at least double the maximum frequency (F_{max}). The formula of the Shannon theorem is expressed like this: $F_e \geq 2.F_{max}$.

With neurofeedback, to obtain exact information about the brain's activity, we need "only" take 256 measurements per second and repeat the measurement over several minutes. Therefore, one session of *classic or dynamic neurofeedback* lasts about 30 minutes. We have precise information about the brain's electrical activity in the spots where the electrodes are positioned thanks to these repeated measurements throughout the session.

Neural activity is not cyclical – unlike tidal phenomena– and measurements must be repeated each session. We cannot make a schedule for the brain!

2.3. Recording the signal

"I want to record the chirp of a cricket but there is always noise from cars or planes, dogs barking, conversations and horns, even distant ones. Any ideas, Jack?"

"Take two microphones – unidirectional is best, because they will not record too many different sounds – and position them like this: one as close as possible to the sound you want to record, and one further away."

"And then?"

"With a sound processing software, we can analyze your recordings and work on isolating the different sound sources."

"How does that work?"

"We have two microphones that record two different sounds and some same sounds. One records the cricket and any sounds in the surrounding area, and the other records sounds from the surrounding area. We mix the sounds and remove what is common to both recordings, therefore all of the identical frequencies."

"With an audio equalizer?"

"No, with sound processing software. First, we have to break down and digitize the sound. We can then decrease the sound volume of the frequencies that we want to remove."

"Can we compare the two recordings?"

"Yes, by overlaying them. This is called 'averaging', and it is used to eliminate 'noise' that is considered as an interference in order to isolate a weak signal."

"Let's try it!"

The microphones to record the cricket and the electrodes positioned on a person's head to measure the EEG have the same function. They pick up the waves that they encounter. For one, these are vibrations in the air (sound waves), and for the other, these are electrical pulses (brain waves).

The difference between the signals picked up by two electrodes provides information about the electrical activity received while also eliminating the "noise" (interference) common to all the electrodes placed on the scalp. The elimination of the "noise" when processing sound occurs through *averaging*[4]. As a result, we obtain information that is as close as possible to the electrical activity produced where the electrodes are located. The electrodes are in pairs: one represents a positive pole (+) and is said to be "active", and the other represents a negative pole (−) and is said to be "passive". They sense the brain waves where they are positioned as well as any waves emitted in the surrounding environment.

It is the difference between two recordings that makes it possible to obtain information that is as close as possible to what we hoped to isolate.

---

4 Averaging involves processing signals by summation: it consists in overlaying signals in order to make the strongest intensities, or interference, stand out against the weaker signal that we want to record. The summation of the two signals will double their intensity and identify the interference. To retain only the weak signal emitted by the brain, the strong intensity signals produced by the environment are eliminated by subtraction.

Aside from averaging, there is another tool that makes it possible to amplify weak signals emitted by the brain: a compressor. This is a processor that suppresses signal waves that are too intense and amplifies waves that are too low. When the intensity threshold determined by the sound technician (or neurofeedback software) is exceeded, the compressor takes action and reduces how the strongest waves are perceived, making the weaker waves more apparent.

"I recorded a conversation, but at a certain point, one of the speakers starts to yell. We need to lower the sound suddenly so we don't get our ears hurt. Is there a simpler way?"

"With the recording of a conversation, in dynamic mode, we can set a compressor at a pitch of 50 dB (decibels). Most of the time, the human voice varies between 20 and 50 dB. If one of the speakers yells suddenly, the voice will abruptly exceed the 50 dB. The compressor reduces the intensity of how it is perceived by the recording device in order to avoid deafening listeners. This way, there is no sudden increase in the sound."

There are two possibilities for signal processing in *dynamic neurofeedback*.

There are five sensors: two active electrodes, noted as (+), in C3 and C4 and three clips on the ears including two passive ones, noted as (−), in A1 and A2 to record waves related to the environment, and one neutral clip, noted as 0, in A2 as well.

The summation of the signals recorded in C3 and sensed in A1 makes it possible to obtain the electrical signal produced by the brain where the electrode is positioned (in C3). The electrode placed in C3 records brain activity as well as signals emitted in the surrounding environment.

The sensor placed on the ear records signals from the environment. Using averaging (the "subtraction" of two pieces of information), we obtain the signal emitted by the brain alone by eliminating the "noise". This mode of calculation is called "unipolar". By doing the same with the electrode and the sensor placed in C4 and A2, we obtain the signal of the brain's electrical activity in C4. This way, we get precise information about brain activity in the left hemisphere in C3, and in the right hemisphere in C4 (Figure 2.4).

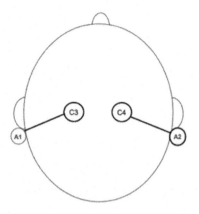

On the screen, " C3 – A1" and "C4 – A2" are displayed as two andependent images of the brain activity in C3 and C4 in each hemisphere. The feedback occurs on two separate signals.

Figure 2.4. *The signals are processed separately*

To obtain information about the electrical activity in zones C3 and C4 at the same time, the two signals are superimposed and the difference is processed and analyzed by the software. In doing so, we obtain by comparison information about the balance of the two hemispheres (Figure 2.5).

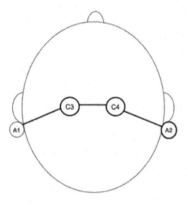

"(C3 – A1) – (C4 – A2)" The remaining signal provides information about the differences between the two hemispheres by comparison. The differential between the two signals informs us about the brain's general activity. The feedback occurs on the difference between the two signals.

Figure 2.5. *The signals are processed together by subtraction*

In *classic neurofeedback*, there are three to five electrodes, or even more: one or two active electrodes, noted as (+), one or two passive electrodes, noted as (−), and one neutral electrode, noted as 0, positioned in A2. It should be noted that there are no clips on the ears with this system.

The signal only undergoes differential processing: the difference (subtraction) between the signals picked up by two electrodes is analyzed and processed, thus eliminating all interference from the surrounding area as well as what is common to each signal (Figure 2.6).

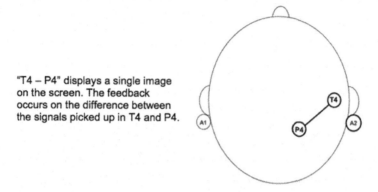

"T4 – P4" displays a single image on the screen. The feedback occurs on the difference between the signals picked up in T4 and P4.

Figure 2.6. *The signals undergo differential processing*

Only the differences between signals recorded by each electrode are taken into consideration in *classic neurofeedback*. This mode of calculation is called "bipolar".

On the other hand, if we position five electrodes, the system can process two signals separately (Figure 2.7).

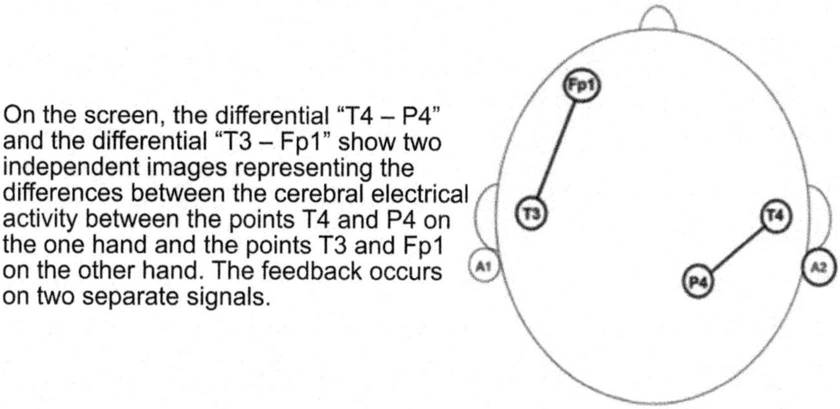

On the screen, the differential "T4 – P4" and the differential "T3 – Fp1" show two independent images representing the differences between the cerebral electrical activity between the points T4 and P4 on the one hand and the points T3 and Fp1 on the other hand. The feedback occurs on two separate signals.

Figure 2.7. *The signals are processed separately two by two*

In this matter, the difference between the two systems is that *dynamic neurofeedback* considers the entirety of the signal emitted by the brain at the point where the electrode is positioned, comparing it to itself over the course of the session, while *classic neurofeedback* prioritizes processing the differences in the variation of the signal between two points, thereby comparing the two signals to each other and to a reference standard developed from the patient's own electrical brain activity (see section 4.2).

2.4. Signal quality

In order for the EEG analysis by the software to be precise and the neurofeedback to be effective, it is essential to obtain the clearest and purest signal possible. That is why the software also analyzes the quality of the signal. Frequencies whose intensity varies very little or not at all during measurements will be subtracted from the information and considered to be "noise".

Only the larger variations in the brain's electrical activity "at rest" are considered, because they represent a possible dysfunction.

Some frequencies are not taken into account by the neurofeedback system. These are the so-called "interference" frequencies produced in the surrounding environment: primarily electromagnetic waves, emitted at 50 Hz in Europe (60 Hz in English-speaking countries). These waves are to be found equally in the brain and the environment. They are not taken into consideration directly by the neurofeedback software and are filtered.

Other frequencies are not necessarily analyzed by some software, including frequencies known as "*Schumann resonances*"[5]. These frequencies correspond to the Earth's electromagnetic field, the vibrations present between the ionosphere and the surface of the Earth, whose intensity varies according to weather conditions, the alternation of day and night, and

---

5 Winfried Otto Schumann: German physicist (1888–1974). Schumann resonances are a set of infra-low frequencies (from 7 to 35 Hz) in the Earth's electromagnetic field. These waves, which are present in the space between the Earth's surface and the ionosphere, are activated by electrical discharges produced during storms. The fundamental wave varies between 7 and 10 Hz (oscillating most frequently around 7.8 Hz). The resulting harmonics oscillate between 14–20 Hz and 21–30 Hz, most frequently around 14.3 and 20.8 Hz, respectively.

geographical location. A storm increases the ionization of the atmosphere considerably and consequently increases the intensity of these frequencies. These frequencies correspond to 7.8 Hz, called Earth vibration frequencies, and 14.3 Hz and 20.8 Hz are harmonics.

These resonance frequencies seem to have changed in recent years, no doubt due to the hyper-electromagnetization of our environment by new technologies. This wave pollution is called *electrosmog* and can affect people who are particularly sensitive to waves ("electro-sensitive").

With neurofeedback, after eliminating the "interference" waves, the difference between the two signals recorded by the two electrodes gives an indication about the frequency variations between the one and the other. This is the basic signal that the neurofeedback system processes, regardless of which system. The signals shared by the two electrodes will be eliminated by the software because they do not present any significant variation in intensity.

Some systems have LEDs (light-emitting diodes) on the amplifier-encoder that makes it possible to verify the clarity of the signal. This is the case in *classic neurofeedback*. Others, like *dynamic neurofeedback*, require a baseline whose analysis gives indications about the quality of the signal.

In all cases, the less clear the signal perceived by the electrodes is, the more difficult reading the EEG will be, and the more imprecise, or even impossible, its processing by the software will be.

> "Look, Marion, I filmed the kids at the beach. With the sun, I couldn't really see what I was filming, but it was pretty funny."

> "Yes, it is rather overexposed! But I would say they had a lot of fun!"

> "I'll turn up the volume a bit so you can hear what they were saying. You're going to laugh!"

> "Strich, schtrouf, rrrrhhh, frfrfrfr, Mommy, schtriff, ffffff, crab, gnegne, schleuch…"

"Ha ha! It's true!"

"Oh no! It was too windy and I wasn't careful! My microphone picked up the wind. The recording is ruined!"

In order for the signal that will be processed to be the "cleanest" possible, it is essential that the recording of information be of good quality.

In order for the electrodes to send the clearest possible information, they must be clean (the client's skin as well), and properly positioned, attached to the scalp with a conductive paste. The feedback, and therefore the effectiveness of the method, depends on the quality of the signal.

2.5. Band-pass filters

Once the signal is received, it is processed and analyzed. The software sends the feedback to the brain, or in other words, returns information about how it functions, in order to signal potential imbalances it has observed. These dysfunctions correspond to variations in intensity that are too great over groups of frequencies determined by the filters. These variations in intensity characterize the "variability" of the signal.

Electrical activity in the brain is broken down by the software into frequencies from 1 to 40-42 Hz. The software does not send feedback about the entirety of this electrical activity, but about its variability over time.

Below is a simulated signal (Figure 2.8) created for the purposes of this explanation.

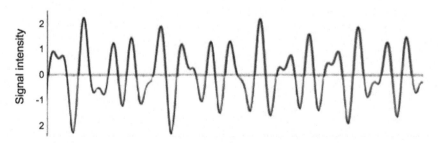

Figure 2.8. *The simulated signal composed of four superimposed frequencies*

In addition, here is the same signal broken down into frequencies by the Fourier transform (Figure 2.9). We can see four different intensity frequencies.

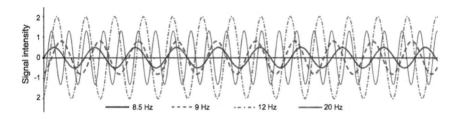

Figure 2.9. *Breakdown of the previous signal into isolated frequencies. For a color version of this figure, see www.iste.co.uk/vincent/neurofeedback.zip*

With only four frequencies, the information is already hardly useful. It would be difficult to work with 42 frequencies at the same time, with different intensities and phases.

The spectrum of a signal – its breakdown into frequencies – is easier to use if it is presented in the form of a frequency histogram.

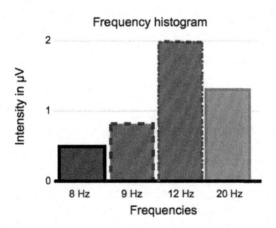

Figure 2.10. *Spectrum of the previous signal, broken down into a frequency histogram. For a color version of this figure, see www.iste.co.uk/vincent/neurofeedback.zip*

In order to better target imbalances and therefore be more efficient with feedback, the software is equipped with filters. They are called "band-pass filters" and they make it possible to select the frequencies whose variability is relevant for processing the signal.

"Band-pass" filters are designed to target brain waves based on their frequencies, like a sound equalizer makes it possible to modulate the treble and bass of an audio recording. For each filter, the activation thresholds for the feedback are recalculated according to the intensity and variability of the waves identified in this way.

We often use filters in several domains in our daily lives. In sound processing, especially, but not exclusively:

"Hello? It's Katie!"

"Hello, Katie. How are you?"

"Good, and you? I edited the photo, I'm sending it to you by email. Tell me what you think. I included it in the mock-up for your information brochure so you have an idea what it would look like."

"I'll take a look right way. Thanks for doing it so quickly! I have it up on my screen now. The rendering is good, it's dynamic and soothing at the same time."

"Like we said, I integrated the photo in the scene. Can you see it? I cropped the silhouette of the woman and slightly modified the color of her sweater to match the colors in the mock-up."

"How did you do that?"

"With image processing software that creates graphic layouts through the layering of layers."

"Can you explain that to me?"

"In computing, digital image processing follows the same rules as sound processing, where filters are used. For pictures, these filters are called 'masks' and they make it possible to work with and modify the transparency of different layers. We can make changes to the clarity, color temperature, shade, contrast, etc. to obtain a cohesive result that is appropriate for the context."

3

The Principle of Homeostasis

The similarities between sound processing software and neurofeedback software are explained by the fact that the signal emitted by the brain can be translated into frequencies, like sound.

A sound is a vibration in the air and electrical activity in the brain is an electrical pulse that follows the same mathematical laws as sound, in that it is represented by a sinusoidal curve that is called a "wave".

That is why we talk about brain waves when we describe the brain's electrical activity. There are different sounds in nature: bird songs, animal calls and noises created by humans, like music, in particular. In the brain, brain waves also differ based on their location and the nature of the neurons that transmit them.

It is essential that the brain's electrical activity maintains a harmonious balance so that all of its capacities are operational and able to be mobilized when needed, like instruments in an orchestra are in tune so that the music is as harmonious as possible.

3.1. Brain waves

Brain waves are very low frequencies and vary between 0.1 and 128 Hz.

They are classified into groups that bear the names of Greek letters. The waves can be delta (0.1–4 Hz), theta (4–7 Hz), alpha (8–13 Hz), "low" beta waves (13–20 Hz) including SMR (sensorimotor rhythm) waves

(13 Hz–16 Hz), "high" beta waves (21–40 Hz) and gamma (40 Hz and above). Alpha, beta and gamma waves were discovered before delta and theta waves, which explains why the classification of brain waves is not in alphabetical order.

When we set a band-pass filter, we consider the intensity of the signal emitted by certain waves included between the limits of the filter and we remove the signal of other waves outside of these limits. The software measures the variation of the signal on each frequency (Figure 3.1). If the frequency has too much variation, it represents an instability.

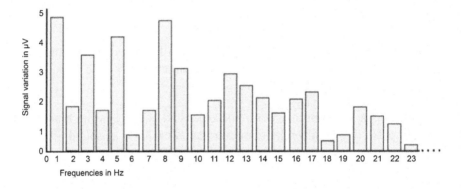

Figure 3.1. *Histogram of the variation of frequencies from 0.1 to 23 Hz measured over 1 second*

As shown in Figure 3.1, the frequencies 6, 18, 19 and 23 Hz have a very low variation (lower than 1). This means that the intensity of these frequencies over the course of measuring the signal (256 times per second) varied at a mean rise of less than 1 µV.

Setting a filter is the equivalent of listening to a certain group of instruments in an orchestra and ignoring the melody played by the others, like listening to the string instruments without listening to the wind or brass instruments at the same time.

In order to target only a single group of frequencies, we must multiply the signal of the frequencies in question by 1 and the signal emitted by all other frequencies by 0 (Figure 3.2). The software will then only consider information concerning the targeted frequencies.

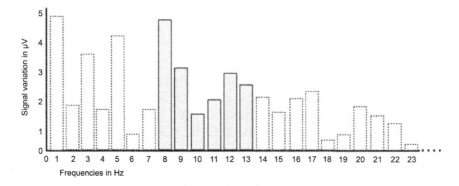

Figure 3.2. *Filter placed on alpha waves*

There are "band-pass" filters to filter delta waves, theta waves, alpha waves and so on. These filters play a major role in neurofeedback, because each frequency group corresponds to a mental state of the individual, to brain activity, that is directly correlated to different parts of the brain.

With *classic neurofeedback*, there is a module that has alpha and theta band-pass filters in which the rest of the electrical brain activity is not considered. When we set these two filters, only waves between 4 Hz and 13 Hz are analyzed. This makes it possible to target one electrical activity in particular, by focusing the training on brain waves involved in fear and anxiety, primarily from 5 Hz to 10 Hz.

"Brain waves" are the simultaneous activation of several thousands of neurons indicating a specific activity of the brain. The information circulates from neuron to neuron through the production of neuromediators, from one group of neurons to the next until it activates the trigger for the appropriate reaction through the neurotransmitters. When the intensity of a frequency is very high, it means that there are many neurons active on this wavelength, and that this is the dominant activity of the brain at that time during the measurement.

The lack or excess of synchronization of brain activity can be a sign of disorders, such as sensorimotor issues. Similarly, an excess of cerebral synchronization can be an indication of a condition: epilepsy is characterized by too much synchronization within the alpha waves in the brain, and therefore an intensity in the activity of these waves that is much too high, which neurofeedback tends to regulate.

Brain waves are associated with certain activities and with triggering specific neurotransmitters that will, in turn, provoke the secretion of hormones by the organism.

It should be noted that certain neurons, when activating, will trigger the production of a neurohormone whose role is precisely to regulate the activity of the neurons that trigger this neurohormone. This produces a system of self-regulation through activating/blocking the system (Figure 3.3).

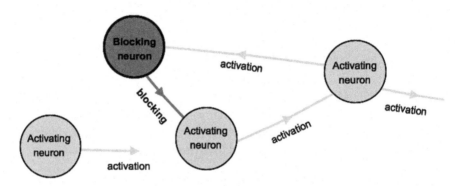

Figure 3.3. *Loop of self-regulation through activation/blocking. For a color version of this figure, see www.iste.co.uk/vincent/neurofeedback.zip*

3.2. The autonomic nervous system

"Can you stay with the two donkeys?" Pauline asked me. "Louis and I are taking Vitamin and Nervous the pony to the woods. There is not enough grass for them here."

"Sure, no problem."

"[...]Because we have to take the road, it's pretty dangerous with the bend that leads right to the field's gate," she added. "We will come back to get the donkeys next."

"The pony is still just as wild and if we take them all at the same time, they'll be frightened by any car that comes along," said Louis.

"It's true, cars tend to take that bend very fast! And Nervous was well-named!"

"To get there, we could cross the stream, but it is too steep. This winter, the mare fell into it. It was pouring! The stream flooded and she drowned."

"Louis nearly got killed as well trying to pull her out."

I watched them move away down the road, leading the horses by the harness.

The forest wasn't far, I could see it from where I stood. It was just the other side of the stream bordered by an electric fence, to avoid any other mishaps.

I held one of the donkeys, Pepita, by the harness while waiting for her turn to change fields.

The horses moved further away. The donkey started to bray and became upset, showing signs of stress and pulling on her tether.

She brayed and looked desperately toward the horses which were moving further away.

"Calm down! They'll come back to get you. Be patient. Quiet!"

I gently stroked Pepita to reassure her, but she didn't know me well. My voice did not calm her down.

Suddenly, she violently pulled on her halter and got away from me. She galloped in the direction of the horses that she could see in the woods on the other side of the stream.

In one leap, she jumped over the electric fence and skidded to a stop at the edge of the sharp drop down to the stream. She brayed even louder, running up and down the length of the stream. She looked extremely frightened!

I did not understand this fear. The horses had not disappeared, they were close by and she could see them.

Louis and Pauline returned. They caught the frightened donkey, not without some difficulty. The route they had to take temporarily led them away from the horses and the donkey believed she was separated from them. She resisted.

"They have been together since they were very small, you see?" said Pauline. "Vitamin is like a mother to her. She thought she had been separated."

Louis was intrigued:

"I don't understand how she was able to leap over the fence! It's supposed to be a psychological barrier."

"Yes, but she did! And at the risk of falling in the stream herself!"

The brain sends "ascending" signals from the subcortical part to the cortex and descending signals toward the rest of the body through neurotransmitters. In other words, when the autonomic system is not "calm", it sends stress signals toward the rest of the brain, which can trigger its blocking. This could explain why a child who is hungry, tired or troubled may not perform well at school. The child has trouble concentrating. His subcortical brain needs to be calmed.

The subcortical brain – also known as the "archaic" or "reptilian" brain – is dedicated to processing automatic behaviors, reflexes and homeostasis in the body through the sympathetic and parasympathetic central nervous systems. This means that its function is to maintain breathing, body temperature, blood sugar levels, etc., at a stable equilibrium, and consequently also manage the triggers of hunger, thirst and fatigue that, when they are satisfied or alleviated, provide the organism with the provisions it needs: nutrients, vitamins, water, rest, etc.

The autonomic nervous system is also responsible for an individual's instinctive reactions when faced with an exterior event. This is called the "survival instinct".

This survival instinct does not obey a process of reasoning, but rather a protective reflex that pushes the individual to react when faced with a potential exterior danger. The possible reactions activated by the subcortical brain are fight, flight or expectancy. These reactions have origins in associated instinctive emotions: anger (fight), fear (flight) and curiosity (expectation).

When the instinctive emotion is too strong, the central nervous system sends signals that are too intense and the associated reflexes can lead to inappropriate, or even dangerous, behaviors.

In the case of the donkey Pepita, the fear was too strong and it pushed her to jump over the electric fence which, until that point, had represented an "insurmountable" boundary beyond which there was a danger. Here, we are talking about exceeding homeostatic thresholds.

Neurofeedback prompts the autonomic nervous system to regulate the production of the neurotransmitters associated with it: acetylcholine, produced by the parasympathetic system as a blocker, which is responsible for the general slowdown of the organism, and adrenaline and noradrenaline, produced by the parasympathetic system as an activator, which are responsible for accelerating the heart rate and breathing rate, among others.

The role of neurofeedback is to train the brain to maintain itself within the thresholds of safety and consequently regulate the reaction of the central nervous system.

3.3. Homeostasis

"Nicolas is burning up with fever! I will take his temperature. 40.1°C! That is much too high! He is shivering, all of his limbs are trembling. This cannot go on, we must lower the fever."

It is after midnight on Saturday night ...

This often happens with children. They fall ill on the weekend. Who knows why.

I looked in our little medicine cabinet... Nothing. Well, nothing that could help. There was no more acetaminophen, no aspirin (I'm allergic to it, so we don't keep any in the house. You never know, accidents happen so fast!)

In any case, he doesn't know how to take pills and when he drinks, he vomits. As for suppositories, pardon the expression, but he... "spits" them out.

What to do?

I hesitated to call the doctor on duty. He was often swamped and this was probably not an emergency. But all the same, I was not calm.

When I was a child, I often had a fairly strong fever, like this, and my mother put cold compresses on my chest. She wrapped my wrists and ankles in washcloths soaked in cold water... which seemed freezing to me!

The old remedies had proven themselves.

I put compresses on Nicolas' chest, feet and hands. He talked to me. I didn't understand any of it. He was delirious. The fever was too high.

I dipped the compresses in cold water. They quickly heated up again when they came in contact with his feverish little body.

After fifteen minutes of this treatment, I took his temperature again. 39.8°C. I gave Nicolas a few mouthfuls of ice water.

I repeated the process an hour later.

The next day, his fever was lower. It was only 38.8°C. A few more compresses throughout the day and it had totally disappeared.

In situations like this, certain maintenance thresholds of the body's balance are exceeded: increase in body temperature, for example, as well as gradual dehydration. If this situation carries on for several days, it is dangerous. We cannot go on too long with too high of a temperature, especially a child.

Homeostatic thresholds represent the limits of an interval in which the body does not run any risk. If these limits are exceeded, the organism initiates a process of self-regulation in order to return to this interval of safety.

If these limits are exceeded too long and too often, the organism suffers, which can lead to issues that can sometimes be very severe: diabetes, scurvy, heart attack, etc.

The brain also has homeostatic thresholds beyond which it suffers. An excess of alcohol can lead to *delirium tremens*, even an alcoholic coma, severe nutritional deficiencies or excesses, and to brain disorders.

An excess of stress, a lack of sleep or emotions that are too strong inevitably lead the brain outside of its homeostatic thresholds. The brain can, by itself and with its own mechanisms, regulate itself and maintain a stable balance, but when it is no longer able to, it needs help to do so.

A stable equilibrium is essential for the proper functioning of organisms and their brains as well as their survival.

3.4. A stable (but variable) equilibrium

If we recognize that there are tolerance thresholds common to all human beings, below and above which the organism runs a risk, it is also understandable that these homeostatic thresholds can vary depending on age, height, weight, physical condition, etc., as can the organism's ability to resist vary from one person to the next.

A nursing infant cannot resist dehydration very long. The limits of its safety thresholds in this area are much lower than those of an adult.

An Inuit has a greater resistance to the cold than a person from a country with a warm climate, for example, and conversely with heat.

However, it is possible to actively and voluntarily regulate the capacity of an organism to maintain itself within its comfort range. A person who has never practiced a sport will, when exercising, have a less developed muscular, respiratory or cardiac resistance than a high-level athlete. This person will more quickly exceed his thresholds of stability and balance.

On the other hand, the high-level athlete who trains several hours a day will have a better stability and a higher resistance to the effort than the "weekend warrior" who does not have the physical resistance to run a very long distance on their first outing of the season.

Nevertheless, the occasional athlete can, if he makes the choice and dedicates the necessary time and energy, become a long-distance runner (with some physical conditions first, because after all, not all humans are the same). By training, his physical and sports performances will improve.

This means that it is possible to modify an organism's capacity to maintain a stable equilibrium. It is also possible to extend one's own homeostatic limits.

4

How Neurofeedback Works

We have seen that the organism is capable of regulating itself spontaneously and quickly when circumstances require it (flight for survival) or slowly when there is no imminent danger (adaptation to a new environment).

We have also seen that it is possible to act on this equilibrium through regular, sensible, controlled and appropriate training.

The brain can also be trained to regulate its operation and optimize its performance.

Classic neurofeedback uses "band-pass" filters, including a "frequency band-pass" filter. The brain waves associated with these low frequencies correspond to information directly related to the subcortical brain responsible for automatic responses, reflexes and basic survival emotions like fear and anger. When an individual is scared, the signals sent to different parts of the body are emitted by this archaic brain.

During a neurofeedback session, the band-pass filters are activated. The EEG is analyzed and the information is processed. When the variation of the intensity of a frequency (or group of frequencies) is too irregular in the range targeted by the band-pass filters, the software sends feedback to the brain. This feedback is information correlated to neural activity, and perceived as a strong signal by it, because it is interpreted as a potential, albeit fictitious, danger.

The brain will attempt to react to this virtual danger by implementing a self-preservation strategy.

4.1. The orchestra conductor

Feedback sends the brain instantaneous information about its own operation when the homeostatic thresholds of its electrical activity are exceeded. These thresholds are determined for each person and are never predefined or standardized. The feedback is adapted to frequency variations in the patient's brain (Figure 4.1).

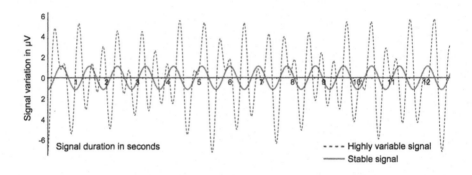

Figure 4.1. *Two electrical signals recorded for 12 seconds. For a color version of this figure, see www.iste.co.uk/vincent/neurofeedback.zip*

By recording the electrical activity, we detect all of the brain's homeostatic thresholds, which is to say the limits that its global electrical activity rarely, or never, exceeds.

By breaking down the brain's electrical activity into frequencies, we obtain the equivalent of the score of every musical instrument in an orchestra, which the orchestra conductor must read and from which he will generate the modifications to be made. This score is represented by a frequency histogram on which we can distinguish the delta, theta, alpha and beta wave groups represented by different colors in Figure 4.2. This type of 2D (two-dimensional) histogram is used in *dynamic neurofeedback*.

The neurofeedback software plays the role of the orchestra conductor who will indicate to the brain-orchestra the corrections to make to the rhythm, intensity and cohesion of its activity so that the whole thing is harmonious.

It is important to understand that feedback cannot be standardized: it is directly related to the electrical activity proper to each person's brain. A person's brain activity is unique and the "localized" EEG – recorded by the software through the electrodes – corresponds to the brain activity of that person, which cannot be compared to any model.

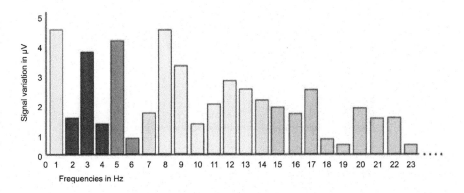

Figure 4.2. *Histogram of frequencies from 0.1 to 23 Hz in 2D. For a color version of this figure, see www.iste.co.uk/vincent/neurofeedback.zip*

The "software orchestra conductor" receives the "histogram-score" of the personal "electrical activity symphony" of the person having a session, then analyzes and detects the "intensity variations-disturbances". Most of the time, these disturbances are due to too much variability in the electrical signal, indicating an instability on one frequency or another (musicians who play too loudly or too quickly compared to other musicians). The excessive oscillation of the intensity measured can trigger feedback.

It should be noted that the histogram displayed on the clinician's screen in *classic neurofeedback* appears in 3D – three dimensions – the third dimension being time (Figure 4.3). This makes it possible to visualize the evolution of the variation of different frequencies over the course of a session and, in doing so, verify whether a peak is infrequent or whether it is reproduced frequently over time, or to see the evolution of the brain's electrical activity during the session, including namely relaxation.

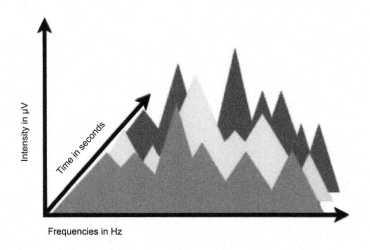

Figure 4.3. *Histogram of frequencies in 3D. For a color version of this figure, see www.iste.co.uk/vincent/neurofeedback.zip*

4.2. Calculating thresholds

A thermometer indicates a temperature that differs depending on the reference system. For example, 20°C seems like a nice and warm day after the winter. For the temperature of a cup of coffee, it is rather cold. For the human body, this temperature is fatal.

Homeostatic thresholds differ from one person to the next (see section 3.3), but they do have common characteristics. Even if some people have enough endurance to withstand extreme stress (lack of oxygen, very low outdoor temperatures, lack of food), most people cannot survive for long outside of their individual thresholds.

We can therefore assume that there is a threshold of comfort between the limits of which the person does not run any risk. Depending on the mode of calculation, this threshold varies for each factor.

If we consider the brain's electrical activity, the signal recorded by the encoder is analyzed and broken down into frequencies represented by a histogram (see section 2.5). This histogram considers two factors related to the brain's electrical activity: the frequencies and their variation over time.

Dynamic neurofeedback processes the signal emitted by the electrical activity in the brain dynamically, which means that each event is compared to the previous one and each measurement is compared to the previous one.

The signal is processed at the time "t" and then "t+1", and so on (Figure 4.4).

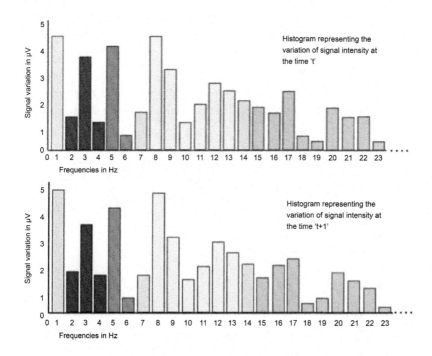

Figure 4.4. *Histograms of dynamic neurofeedback at times "t" and "t+1".*
For a color version of this figure, see www.iste.co.uk/vincent/neurofeedback.zip

This comes back to the idea of measuring the tides several times each day for a year, not considering the pre-determined schedules. It is an instantaneous measurement compared to the previous ones. If there is a storm, the measurements before and after this weather disturbance will be very different and translate into a momentary imbalance in the system.

With *dynamic neurofeedback*, this is equivalent to measuring the brain's electrical activity 256 times per second throughout the duration of the session (about 30 minutes) and comparing each measurement to the previous one. If the signal remains stable, this represents a histogram with little variation in the frequency intensity in the brain. An increase in the frequency variation range indicates the appearance of a possible disturbance.

Each signal variation will be analyzed and, if it is repeated and increases too frequently, the feedback will be triggered in order to prompt a return to stability.

With this system, there is no model with which to compare the electrical activity in the brain. The software uses a dynamic histogram indicating at each period the measurements of the signal state and the intensity of each frequency. This is not, of course, 256 variations of the histogram per second, but the periodic representation of the signal variation every tenth of a second, for example. The superimposition of the graphical representation of these periodic measurements gives the visual illusion of a dynamic, moving histogram.

If we superimpose the two events in the previous histogram, we obtain a difference in frequency intensity between the times "t" and "t+1". The succession of measurements in the following times, "t+2", "t+3", etc., gives this illusion of movement, visible on the screen of the *dynamic neurofeedback* practitioner (Figure 4.5).

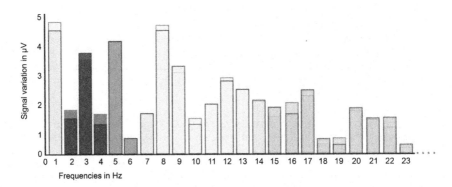

Figure 4.5. *Superimposition of two histograms at times "t" and "t+1" showing variations of signal intensity. For a color version of this figure, see www.iste.co.uk/vincent/neurofeedback.zip*

In the *classic neurofeedback* system, the histogram is in 3D – three dimensions – and also considers changes in time (see section 4.1). The persistence on the screen of intensity variations (from 5 to 10 seconds) makes it possible to view the intensity peaks and their duration, signaling a possible anomaly. The signal is processed based on normal ranges measured by the software at the time of the baseline during the first seconds of recording brain activity. The signal processing uses the normal distribution or *Gauss's law*.[1]

In the case of thresholds that trigger feedback, in relation to normal ranges, the variations in the brain's electrical activity are measured 256 times per second. The calculation of feedback thresholds represents a Gaussian curve in which the normal range constitutes the brain's space of comfort, and whose limits, plus or minus 10%, constitute the homeostatic threshold tolerance (Figure 4.6).

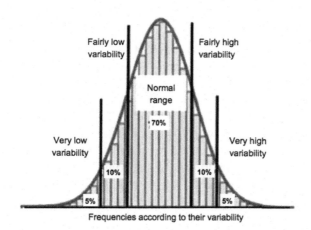

Figure 4.6. *Gaussian curve presenting the upper and lower thresholds that trigger feedback, as well as the normal range triggering the reward. For a color version of this figure, see www.iste.co.uk/vincent/neurofeedback.zip*

---

1 Carl Friedrich Gauss: German mathematician, astronomer and physicist (1777–1855). The normal distribution or Gauss's law is one of the laws of probability that is most adapted to model natural phenomena resulting from several random events. The Gaussian curve has an essential place in the study of random phenomena thanks to the central limit theorem. This theorem says that if we average the measurements taken for a single phenomenon, the values observed are similar to a Gaussian curve as soon as these measurements are numerous enough. This result does not depend on the nature of the phenomenon observed, but on the condition that the measurements are independent of one another.

The thresholds that trigger feedback are calculated from the normal ranges. Feedback is sent when the variability of the electrical activity exceeds the tolerance thresholds, as long as the system does not return to the normal range.

The reward system in the *classic neurofeedback* system (see section 1.2) is initiated when the brain's electrical activity is maintained within the normal range and calculated in relation to the person's brain.

4.3. The reward

DEFINITION.– The reward only exists in the *classic neurofeedback* system. It is a signal indicating that the brain produced the correct reaction and regulated itself. When the reward is not produced, it signifies that the brain is in the process of looking for a solution to regulate itself that it has not yet identified.

Nevertheless, certain sessions may not use rewards, like the alpha-theta module described earlier (see section 3.1). The reward is a signal that serves to anchor the modifications in the neural system: it is a system of reinforcement.

On the screen, *classic neurofeedback* clinicians have indicators that allow them to see if the brain tends toward regulating itself. The clinician can see if the brain is approaching the threshold that triggers the reward, indicating to the person that the brain activity is being maintained within the limits of the homeostatic thresholds (Figure 4.7).

Patients can attempt to maintain this balance through biofeedback exercises whose results are displayed on the client screen, indicating the effects of their attempts to self-regulate. It should be noted that the activation thresholds can be modified in order to increase or decrease the difficulty of the brain training during the sessions, in order to gradually optimize the associated cerebral functions and adapt the training to the person's sensibility, as some people can become anxious when they do not immediately obtain a satisfactory response to the request even though most of the training occurs on a subconscious level.

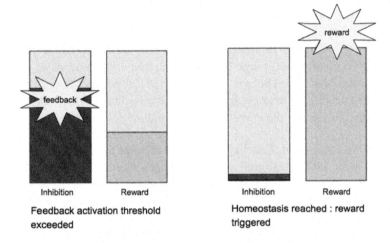

Feedback activation threshold exceeded

Homeostasis reached : reward triggered

Figure 4.7. *Markers indicating the thresholds that trigger feedback and reward. For a color version of this figure, see www.iste.co.uk/vincent/neurofeedback.zip*

The calculation of activation thresholds for feedback and rewards does not only consider the mean electrical activity of frequencies in the brain. It also takes into account the *standard deviation*[2] that gives an indication of the distribution of these intensities.

Theo and Charlie came home from school with their report cards. They compared their grades and their assessment in math.

– "12/20. Solid student. Consistent and regular results. Keep up the good work," read Theo. That's encouraging!

– "15/20. Good work overall, but lacks discipline. Results are not consistent. Take learning more seriously," read Charlie. "I'm better at math than you, but did you see what the teacher wrote about me?"

How many students have, like Charlie, perceived their teacher's comments as unfair criticism? And yet...

2 The standard deviation measures the distribution of a set of data. The weaker it is, the more the values are grouped around the average.

Let's take a look at their grades. There were six assignments over the term. Here are Theo's grades:

He got: 11.5, 13, 10.5, 12.5, 11.5 and 13. We can see that they vary very little and range between 10.5 and 13 compared to his average of 12/20. There is a difference of 2.5 points between his worst grade and his best grade. The distribution of his grades is fairly consistent: the corresponding standard deviation is low.

On the other hand, Charlie has a better average than Theo: 15/20, but with highly variable grades that indicate, according to the teacher, a lack of seriousness in his efforts at school.

Charlie got: 20, 10, 14, 19, 9 and 18. His average is 15/20, but he had a difference of 11 points between his best grade and his worst grade. The distribution of his grades is heterogeneous: the corresponding standard deviation is high.

Regarding brain activity, the calculation of activation thresholds for rewards and feedback is made using the average intensity of a frequency, the variation of this intensity, the distribution of the intensity peaks, their variability, the recurrence of these variations, etc. These factors are all considered by neurofeedback software.

4.4. Feedback

There are several types of feedback: sensory (auditory, visual, tactile) and biofeedback (not all neurofeedback systems have this category).

Each system has its own feedback. Although every system uses visual and auditory feedback that can be presented as modulations or interruptions of sounds and images, the *classic neurofeedback* system also uses visual, auditory and tactile rewards as reward and reinforcement systems. Sounds, lights or vibrations indicate to the patient that a balance has been reached. That is how the learning process takes place: by reinforcing a route taken again and again.

Biofeedback is information about the heartbeat, breathing, body temperature or skin impedance. This information is gathered by sensors. Once the person has feedback about his biological indicators, he can gradually learn to modify them, thereby developing regulating strategies whose effects are immediately measured by the sensors, indicating whether the intended solution is appropriate or not. This is the difference between the *dynamic system* (without biofeedback) and the *classic system* (with biofeedback).

It should be noted that it is entirely possible to complement a system, dynamic or classic, with other biofeedback tools like cardiac coherence software, for example, which represents the patient's heart rate on a screen, measured using a sensor positioned on his ear or finger. In doing so, the patient can attempt to regulate it by practicing breathing exercises whose effects are instantly visible on the screen.

The most important thing is that the patient and his brain have the maximum amount of information about their own activity to be able to regulate themselves: one consciously, the other subconsciously.

Too much variation in the intensity of an electrical frequency in the brain represents an instability that can be correlated to disorders like anxiety, insomnia or depression. The software positions the band-pass filters and determines the intensity thresholds above and below which the balance is unstable according to the global electrical activity of the brain. These are the homeostatic thresholds noted earlier (see section 3.3). They are calculated in direct relation to the brain function of the person participating in the session.

If an orchestra plays a symphony at a slow rhythm, then this rhythm will be taken as the base for the calculations. The variations in this rhythm will be considered to be imbalances that trigger feedback. The rhythm is personal and depends on the individual, as each person plays their own symphony that is unlike any other. Therefore, regardless of the system, there is no predefined feedback threshold.

The feedback activation thresholds are determined by the lowest and highest intensities between which the brain activity is considered to be stable and normal for the person.

Concretely, in *classic neurofeedback*, we record variations in the intensity of the brain's electrical activity (for example, between 8 and 13 Hz with band-pass filters). When this variation is low, the system is considered to be stable and well balanced. When the variation is too great, the divergence recorded signals an instability in the balance of the brain's activity on those frequencies.

The divergence therefore provides information about the deviation between the variations above and below the homeostatic thresholds. The average intensity of the frequencies measured over the course of a session varies between two values. These values are personal and differ from one individual to another. All instances when the threshold is exceeded are considered "abnormal", which triggers the feedback that will persist as long as the range of the signal variation remains too great, indicating a disturbance in how the brain activity functions.

In order to perceive the electrical activity of frequencies with low intensity, *dynamic neurofeedback* uses a compressor (see section 2.3). The thresholds for the positioning of the compressor vary depending on the session.

It should be specified that the threshold will constantly be subjected to micro-variations since it is recalculated according to the variability of the targeted amplitudes during the positioning phase. This variation in intensity is called "emerging variability".

For instance, if the threshold is positioned in order to filter the waves whose emerging variability exceeds 10 µV, and if during this phase of the session this variability decreases, the threshold will be re-calculated according to this decrease in order to maintain the correlation between the brain's activity and the feedback, thereby remaining as close as possible to the brain's activity in real time.

For example, the compressor suppresses the signal emitted by waves above 10 µV, then decreases the compression threshold to 5 µV (Figure 4.8), and so on, during the different phases of the session, in order to perceive the frequencies with the lowest variability.

It should be specified that during the session, although they are positioned to regulate the amplitude of the variability of the brain's electrical activity at a given phase, these thresholds are constantly re-calculated according to this very variability. Since the measurements made throughout the session are compared to each other, if the variability is submitted to significant changes, the threshold which triggers the feedback will be re-calculated immediately in order to take the change into account and to remain as close as possible to the signal. The threshold is then said to be "dynamically" re-calculated in real time.

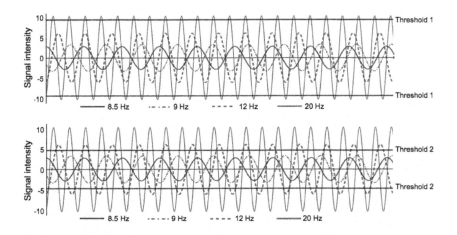

Figure 4.8. *Positioning of the compressor at two instances in a dynamic neurofeedback session. For a color version of this figure, see www.iste.co.uk/vincent/neurofeedback.zip*

In sound processing, this is equivalent to perceiving the voices of people talking quietly by suppressing other voices that exceed a certain decibel threshold (50 dB at the start, then 45 dB, and so on). With a compressor placed at 45 dB, the sounds whose intensity exceeds this threshold are multiplied by 0 by the compressor, which is equivalent to eliminating them. Only the sounds whose intensity is lower than 45 dB are picked up with such a compression threshold.

How long a compression threshold is set depends on the selected session. The lower the threshold, the more we find frequencies with low variability, and the more we approach relatively stable activity.

This encourages maintaining each group of frequencies between the limits of homeostatic thresholds, and also makes it possible to determine the limits for each person at the moment when their brain activity is recorded, for which too much variation will trigger feedback.

5

Training the Brain

A flock of starlings flew over the field sleeping beneath the summer heat. In search of insects and butterflies, the flock darted around, coming and going in an unending, whirling ballet. It drew arabesques on the threatening sky. Like a bluish scarf, it seemed to float, hanging from the storm clouds that towered behind it. The wave curled around itself, dancing and stretching out as if guided by the rustle of the wind passing over the wheat fields. A whisper traveled through the stalks rocked by this caress as if they were also carried along by the music. Time seemed suspended to this pendulum.

A loud bang echoed through the countryside crushed with heat. Time stopped. Nature stood still. The flock of starlings, the only movement in this immobile landscape, broke, fell apart, then just as quickly reformed denser, faster, determined to escape the danger that threatened it.

Another blast shattered the silence. Then another…

The starlings changed course. After a moment of panic, the flock reformed and fled toward the village at a dizzying speed. The wave that traveled across the sky spread and the flock vanished over the hills that formed the horizon.

The storm moved away and the clouds dispersed. The sun returned. Nature took up its calm rhythm once again.

The hail cannons fell silent. The storm had passed.

5.1. The action of feedback

A danger, real (cannon) or fictitious (thunder), provoked the same defense mechanisms for the starlings. A blast prompted a survival reflex, regardless of where it came from or what the real risks were. The reflex area activated automatically, triggering the birds' escape.

The blast is the feedback sent back to the brain. The flock of starlings is similar to the activity of neurons whose signals spread from one to another in brain waves transmitting information to the next one and triggering a reaction.

There is consistency between neurons: the waves do not spread in a random fashion. The signal transmitted takes a specific path to trigger the appropriate reflexes for every situation.

When the blast occurred, the brain was disturbed and its activity changed in order to protect itself from potential danger. The neural connections activated and deactivated in order to find the best defense strategy, the best way to avoid these repeated signals.

The brain, through warning signals, will gradually change how it functions, like the flock of starlings that ended up fleeing the field for the hills, moving away from a place that was deemed dangerous.

These survival strategies operate on a subconscious level, because they are developed by the subcortical brain which is not governed by reasoning or analysis.

The survival instinct pushes the brain to react and the starlings to flee.

Without predators, there is neither cohesion nor reorganization of the starling flock.

Without feedback, there is no reaction or regulation of brain activity.

As we have seen, to process the electrical signals of brain activity, the neurofeedback software transforms it into frequencies from 1 to 40–42 Hz represented on the screen by a graph showing the intensity of this activity on each of the frequencies (Figure 5.1): a frequency histogram in 2D (*dynamic*) or 3D (*classic*).

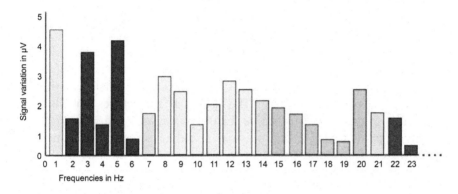

Figure 5.1. *2D frequency histogram with groups of frequencies. For a color version of this figure, see www.iste.co.uk/vincent/neurofeedback.zip*

The "division" of this Fourier transform (see section 2.1) by band-pass filters allows the software to detect the peaks of neural activity on each group of frequencies according to the filter in position.

A feedback activation threshold corresponds to each filter (or group of filters) each time that the electrical activity of a group of target frequencies exceeds the threshold calculated by the software as being the limit of the intensity past which homeostasis is no longer regulated (Figure 5.2). These thresholds are calculated with averages, standard deviations, variation ranges, etc.

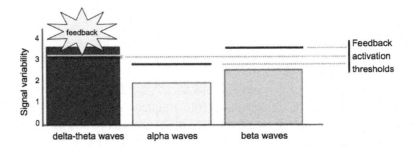

Figure 5.2. *Feedback activation thresholds. For a color version of this figure, see www.iste.co.uk/vincent/neurofeedback.zip*

Feedback provides information about variations in the brain's electrical activity intended for the brain itself. Each activation signals to it that there is something to modify. Each disturbance in the medium (visual or auditory) is correlated to this brain activity. The group of frequencies on which the feedback is triggered is informative for the practitioner and, potentially, for the person participating in the session, revealing what the brain needs to regulate.

> Young 19-year-old M. was very ill. He had attempted suicide not long before and his mother could hardly sleep for fear that he would relapse. During the first session, he cried, and writhed in the chair as though in intense suffering. His mother was very worried. She did not understand neurofeedback and had come on the advice of a work colleague who had recently come out of depression after 23 sessions.
>
> Mother and son both cried. With a nod, I indicated to the mother that she need not worry. This was not the first time that this had happened with patients, I knew, but this was her son and she was scared for him, which was understandable. I asked M.:
>
> "Would you like me to explain what I see on the screen and what is probably happening in your brain?"
>
> He nodded "yes" without saying the word or even opening his eyes.

"Your right hemisphere runs on a perpetual loop on the low frequencies. It is as if it's blocked by emotions that are too strong. On the screen, there are a tremendous number of 3 Hz in this hemisphere, which, from what I deduce, is a sign of a psychological suffering, while the left hemisphere does not indicate any dominant activity on this frequency. As long as the information remains in the right hemisphere, it causes you to suffer. What I'm looking for on the screen is the transfer of information from your right hemisphere to your left hemisphere. This would mean that the brain has started to treat the distressing events by rationalizing them. This would indicate to me that the mourning process has begun. I hope your brain will understand what we need from it!"

The feedback remained mainly on the low frequencies throughout the session. During the first 15 minutes, it was exclusively with the low frequencies in the right hemisphere that it almost systematically exceeded the threshold and altered the auditory and visual media.

I tried to motivate M. to catch on.

"Keep going, M., it should not be long. Hang in there!"

He was still crying. Suddenly, activity in 3 Hz started to dominate in his left hemisphere. The transfer had begun. In my opinion, the mourning process was on its way. I told him:

"That's good, the transfer work is happening. In a few minutes, you will feel relieved. Your brain is in the process of evacuating traumas. Breathe slowly."

Gradually, he was soothed, grew calm and... fell asleep. When he woke up, he said, "I feel good," and smiled at me. His mother cried again, but happy tears this time.

Young M. had 12 neurofeedback sessions and complementary medical care. He regularly sends me updates. He is continuing his studies and, so far, everything is going well.

"I wish you well, M., and thank you for your trust!"

The feedback activated each time the low-frequency threshold was surpassed was a signal for the brain that it must regulate the dominant activity in 3 Hz, but the brain could not understand this information as such. This signal indicates that something is not right, and that there is a disturbance to regulate. It is the correlation between the feedback and the brain's activity that causes it to modify its activity. It is the repetition of this signal every time there is too much variability on a frequency or a group of frequencies that ends up making it react.

> A hunter-gatherer had ventured deep into the forest. He was alone and on the alert. He did not often move so far ahead of his hunting companions, but he had lost their tracks. He moved forward carefully, as he did not know this area.
>
> Suddenly, a branch snapped behind him. He listened. It's probably a deer, he thought, to reassure himself. Another twig snapped, this time to his right. He stopped and listened again. Another snap, on his left, then another. This time, it could not be a deer. Senses on the alert, he expected to see a wild horde or a predator leap out. Fear rose in him. He felt it, and his heart beat faster.
>
> "Don't give in to panic, react!"
>
> Even with the best will in the world, the hunter could no longer remain in the thickets: he seemed to have become a prey. The danger was everywhere, and gradually circling him.
>
> "You must make a decision. Fast!"
>
> Wait, at the risk of seeing the circle tighten around him and leave him no escape route?
>
> Go forward and meet the danger, fighting against these invisible enemies that have surrounded him, at the risk of losing his life?
>
> Flee as fast as he can in the opposite direction of the noises and only stop once out of range?

The aim of this hunter-gatherer was to find his companions and the cave high on the hill where his tribe had taken up residence, where the view of the horizon was clear enough to see any danger coming.

This is the dilemma that confronts the brain when it perceives the feedback sent by the software. These signals have the same effect on the brain as the branches snapping in the forest: their aim is to push the brain to react, prompt it to change its activity, move away from potentially dangerous limits that it strays near too often and that render its balance unstable; to guide it toward the exit of this "psychological forest" where it got lost because it advanced too deeply down a path that disappeared behind it and that it cannot find alone.

In this way, from snap to snap, feedback to feedback, we lead the brain toward the exit of a labyrinth in which it is enclosed. Every piece of feedback from the dynamic and classic systems says, "Wrong way!" and indicates to it to choose another direction and find another solution toward serenity. Each reward in the *classic neurofeedback* system says, "Continue on this path!" and confirms the solutions found.

The deeper and denser the forest, the longer it takes to get out.

The more intense the troubles that a person suffers from, the more time it will probably take to reduce them. The number of sessions required to reduce the symptoms of depression is often much greater than the number to calm stress or anxiety.

5.2. Neuroplasticity

Everything depends on the brain's capacity to react, therefore its capacity to reorganize itself through new connections or by disconnecting certain neural paths. This capacity is called brain plasticity or neuroplasticity.

In the case of young M., the brain activity was so intense in the low frequencies that the feedback was almost exclusively related to that activity. It was the repetition of this feedback correlated with activity peaks that gradually triggered the brain's reaction. However, this reaction was only possible because his brain found a good strategy to regulate itself: an enormous transfer of information from the right hemisphere to the left hemisphere. The modification to the connections in his brain that rapidly led him to feel calm represents an immense cerebral plasticity and ability to adapt.

A mother brought in her 8-year-old daughter, to whom I could not explain how neurofeedback worked, but I hoped to make her understand what we were trying to do together.

"The brain is a house where every room has a purpose: the bedroom for sleeping, the kitchen for cooking food... and one very important room that we tend to forget (some houses don't have one): the attic."

"At our home, the attic is full of junk!"

"That's often the case. From my point of view, in the brain's attic, there are two parts that communicate through a door. A little elf lives on each side. A pink one on the right and blue on the left."

"That's funny!"

"The pink elf must put everything we take up to the attic into trunks (we give him lots of things that will transform into memories over time). He puts it all here and there, as it arrives. He can be a bit disorganized!"

"Like me! Mom is always telling me to clean my room, but I think that's boring!"

"You're a little pink elf, then? The pink elf will give the blue elf what he has managed to identify. He opens the communication door between the two parts and calls for the blue elf (who sometimes doesn't answer right away because he's napping)."

"He's lazy, then!"

"It's possible, but he only works if the pink elf gives him something to put away."

"Does he put things in boxes as well?"

"Or in drawers. With labels. He sticks labels on things. He always knows exactly what is in each drawer. Once he has sorted and put everything in order, the attic is clean: everything is in its place. We are going to put your little elves to work while you watch a cartoon."

"If it's them putting things away and not me, I'd like that. Do you have a cartoon with horses? I love horses!"

A "well-ordered" brain does not have any dominant activity on the low frequencies, especially the 3 Hz, a frequency related to buried trauma and emotionally charged memories. The absence of an activity peak on the 3 Hz indicates that the two hemispheres have integrated the events that have occurred in its life.

A "healthy" brain is dominant in alpha waves "at rest". These are the frequencies associated with relaxation, relief and serenity. When we are uneasy, these frequencies cannot dominate in a harmonious and regular way. There are few, or even no, activity peaks in these frequencies, with the exception of certain associated illnesses (see section 10.3).

5.3. Activating and blocking neurons (simplified explanation)

When we discuss electrical activity in the brain, what are we talking about?

There is no exchange of electrons in neurons. It consists of the propagation of a signal inside the axon – the longest part of the neuron, terminated by synapses that release ions and neurotransmitters that in turn pass the signal on toward another neuron. This signal is spread differently depending on the nature of the neuron.

We will not focus on the detailed structure of neurons, but their function, in a very simplified way in order to understand the role of feedback during a brain training session.

Neurons are immersed in a salty liquid made of, among others, sodium chloride (NaCl). This liquid contains positive ions, Na^+, and negative ions, Cl^-.

The cell walls in organic tissue are permeable and allow water from outside to pass through to the interior of the cell so that the density and the polarity are identical both outside and inside the cell. This phenomenon is called osmosis.

The phenomenon of osmosis does not work like this in nerve cells because they are protected by an impermeable fatty covering called myelin that does not allow the spontaneous exchange of ions between interior and exterior areas.

When a neuron is inactive (or "at rest"), its environment is positively charged with Na^+ ions (and K^+ among others, but we will not go into details), while the exterior environment is negatively charged with Cl^- ions (again, this is a basic explanation). This is a polarized system.

The activation (excitation) or blocking (rest) of a neuron depends on the successive polarizations inside the axon, the body of the neuron.

When a neuron receives an excitation signal, this prompts the successive opening of sodium channels in the myelin that surrounds the axon. Opening these sodium channels will modify the concentration of ions in the axon in order to reach the threshold of excitability. When this threshold is reached, neurotransmitters are released from the synapses and will in turn excite the next neurons they are connected to, and so on.

The neurons activated to trigger the excitation signal causing the sodium channels to open in thousands of successive neurons are called "activator" neurons.

The neurons that stop the spread of the signal are called "inhibitory" neurons.

COMMENT.– The main neurotransmitters involved in the activating and blocking of neurons are called, respectively, glutamic acid (called glutamate) and gamma-butyric acid (called GABA).

The blocking of the neuron to return to a state of rest occurs through sodium pumps that reintroduce the Na^+ ions into the axon to reach their initial state before the excitation.

As the system is a polarized system, opening the channels involves a temporary permeability in the nerve membrane. The exit of Na^+ ions occurs through osmosis as in a tissue cell. It requires no energy expenditure.

On the other hand, activating the sodium pumps in the process of repolarizing the system requires a great deal of energy: about 30% of the energy used by the brain.

This electrical activity of activating/blocking neural networks is the basis of neural computer programming in which a computer tool is able to modify itself according to the actions and reactions generated.

When the action provokes the appropriate response, the system secures the solution validated in this way by a reinforcement loop (Figure 5.3): this is the learning process.

In the brain, this reinforcement loop is accompanied by the activation of neurotransmitters and hormones confirming the solution found: these are first of all hormones called "pleasure" and "reward" hormones, like dopamine and serotonin, for example. The production of these hormones encourages the brain to repeat the action that triggered them in order to rediscover the pleasure associated with the success.

Neurologists Warren McCulloch and Walter Pitts completed the first work about neural networks at the end of the 1950s. They built a mathematical model of the neuron defined as an "object" receiving a flow of information and triggering an output (signal) as soon as a threshold is reached.

This first definition is the basis for artificial intelligence and made it possible to design systems that modify how they work themselves as they learn. Artificial intelligence is present in several domains: voice recognition, financial investment management, flying a fleet of drones, medical diagnostic tools, humanoid robots, etc.

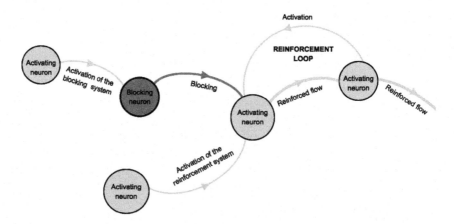

Figure 5.3. *Reinforcement loop. For a color version of this figure, see www.iste.co.uk/vincent/neurofeedback.zip*

5.4. Neurofeedback activators and blockers

When we discuss neurofeedback activators and blockers, we are talking about two different methods:

– *Dynamic neurofeedback* is exclusively *blocking*: the signal is processed in its entirety and the feedback is sent to the brain each time the threshold is exceeded significantly on all frequencies of the brain's electrical activity. With this system, the client is *passive*.

– *Classic neurofeedback activates* and *blocks*. The system can process the signal in its entirety or target specific frequencies in relation to a symptom. The software triggers feedback as long as the brain activity exceeds the homeostatic thresholds and triggers a reward in order to signal to the patient that their response is appropriate. The patient can be *active* and conduct cognitive activities during the session. The software will analyze the brain activity during the exercises.

This means that to block the neural system, the software will only consider thresholds that are surpassed excessively. The feedback will only be activated when there are too many variations in the activity in a group of frequencies.

To engage the neural system, the *classic neurofeedback* software also has a type of feedback called a reward system when the brain is in a state of homeostatic balance.

With each piece of feedback, the brain attempts to understand what is happening. It is a strong signal for the brain, like the blast for the starlings, indicating that there is a potential danger. Just like the hail cannon is a fictitious danger, feedback is a virtual danger (see section 5.1).

"When I heard the bear growl, I started to feel afraid. I didn't immediately understand what it was. The forest seemed so peaceful. There had been some rustling leaves and snapping twigs, but I didn't expect it! When it pounced on John, I was terrified!"

"So was he! He was so absorbed in Mary's wound that he didn't see anything coming."

"Did you see how big the bear was? And its roar after! It could make your blood run cold! And yet it was only a movie!"

The brain nevertheless reacts as if it were real, albeit to a lesser degree because it also gathers information about its environment: a chair, the movie theater, spectators in the room, a neighbor munching on popcorn, and so on, which are all indications that the spectator is not in the scene projected in front of him.

The first reaction of the brain is to trigger the flight reflex process and then, after a quick analysis of the environment, it blocks this reaction and keeps the person sitting quietly on his chair. These processes of activation and blocking occur so quickly that the spectator is not aware of them. They are not voluntary or rationalized. They are formed on a subconscious level.

We rarely see spectators stand up and run out after seeing a bear on the screen, if the brain blocks this information!

This is not the case with children's brains. Their cortex is not yet developed enough to analyze the situation and the reaction of fear manifests itself very strongly in them: shouts, tears and running away.

Movies call upon the capacity of the brain to identify a situation. Fear, frequently triggered by the virtual, is a powerful blocker of the capacities of reflection and judgment.

Advertising and politics have fully grasped this: the former hinges their sales campaign for surveillance system products, for example, on a home's supposed lack of security by capitalizing on their viewers' fears; the latter play on the fear and rejection of what is different to increase their number of supporters.

This is also how neurofeedback works, although there is no message nor product to sell: the feedback is perceived by the central nervous system as a potential danger, which activates a basic and instinctive survival reaction. This reaction in turn generates the blocking, because the feedback is virtual. From one activation-blocking sequence to the next, the brain modifies how it functions, just as the spectator, once tricked by an unexpected scene, will be on his guard when another similar scene occurs, unless too much time has elapsed between the two scenes of this type. The brain will forget its first reaction and reproduce the process of triggering fear and flight.

Dynamic neurofeedback has an impact on a subconscious level: it is no use attempting to control what occurs. This is passive neurofeedback. With hindsight, we can attempt to understand the mechanisms and operations of each person, but the process of regulation occurs independently of a person's will. However, in *classic neurofeedback*, although a major part of the regulation also occurs on a subconscious level, feedback and visual rewards make it possible to optimize the strategies to be implemented in order to improve breathing, for example, and foster relaxation and calm, the results of which are instantly visible on the screen. By concentrating, the user can also become an actor in his own changes: this is active neurofeedback.

The effort of regulating the brain takes training. That is why it is necessary to regularly attend sessions so that the brain does not allow itself to be surprised again by situations that push it to trigger a process that is sometimes inappropriate.

It is only as a result of several training sessions that there are deep changes to how the brain functions. The same goes for physical training that, if it is regular and appropriate, will allow the muscles, lungs and heart to adapt and increase their resistance to the effort, thereby allowing the athlete to improve his performance.

When the feedback intervenes, it is always in direct correlation with the brain's activity when a balance threshold is exceeded on one or more frequencies. With each feedback or reward, the brain regulates itself (or at least, tries to) by activating or deactivating neural circuits which are themselves involved in triggering neurotransmitters and hormones.

The modification of the triggering of neurotransmitters also gradually modifies a person's behavior and sometimes his metabolism.

For example, in the brain, an excess of electrical activity in 3 Hz and 5 Hz seems to correlate with a dysfunction of the production of serotonin, the deficit of which can lead to a state of stress, anxiety and even depression. Through the signals repeated in relation to the peaks of electrical activity in 3 Hz and 5 Hz, the variability of these frequencies and their intensity gradually decrease and the physiological processes associated with them reach a balance (see Appendix 1).

The activation or blocking of neural connections will also, gradually, have an impact on many neurotransmitters and hormones whose production is directly or indirectly connected to the activity of the central nervous system. This is how neurofeedback can foster the regulation of dopamine, melatonin, adrenaline, noradrenaline, serotonin, etc., or regulate blood pressure and blood sugar depending on the brain's plasticity and its own reorganization priorities.

We can hope for improvements in many areas even though they remain entirely dependent on the brain's reorganization priorities and – I must insist on this essential factor – its plasticity.

5.5. Appropriate training

"I thought my lungs were going to burst! I can't do it anymore! I'm quitting!"

"Oh, you know, Antoine, just go at your own pace."

"Yes, but the instructor never stops criticizing us for not going fast enough! Look, there are ten of us in the group; men, women, a few retirees and he's asking us to warm up with twenty laps in the pool with the four swimming strokes! And that's the warm-up!"

"Well, I don't know how to do the butterfly stroke, so I do anything, or rather, I do whatever I want to! Just do the same."

"I'm here to learn how to swim, but the course is not suited for our different levels. It's too hard!"

"It's true. We have already told him, but he decided that he knew best, so we can't do anything."

"Next year, I'm not registering again, or anyway not in his course. What about you, Philip? Shoot, he's coming!"

"Are you napping? Get to work, slackers!"

"He's getting on my nerves!"

"Go on, Antoine, be quiet and swim!"

It's obvious that the swimming lessons proposed are not appropriate for the level of the group. The instructor is not taking his students' needs and abilities into account.

The same goes for neurofeedback: there is no standard session that will suit everyone.

It is essential to adapt it to the person who has come for a consultation. That is why all neurofeedback systems can be modified according to the criteria related to each person (although the modification possibilities are decreasing with the software updates for the current dynamic system, which is, in my opinion, a great pity!).

The training received by neurofeedback practitioners or clinicians is essential in this matter, because understanding the software and its functionalities is an obvious condition to knowing what we are doing and why we are doing it in order to adapt the session to the person we are treating, without which the practitioner is no more than a technician positioning and removing electrodes, without really providing support to the client, who does not in fact know much about what goes on during a session.

Some of the tools available to the practitioner include the different filters that we mentioned earlier, even if most of the time they are set by the software.

The feedback activation thresholds can be modified with filters like a sound equalizer (Figure 5.4) when listening to music.

We can increase or decrease the difficulty of a session by lowering or raising the tolerance thresholds above and below which there will be feedback. For example, if the brain's activity on a given frequency exceeds a certain intensity threshold, there will be feedback to indicate to the brain that the threshold has been exceeded. The thresholds are calculated by the software in relation to the activity that is unique to the person. Sometimes, however, these thresholds prove to be too low (Figure 5.5) and there is too much feedback, which can "stress" the brain.

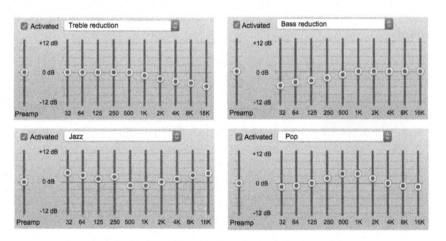

Figure 5.4. *Different sound equalizers, according to the desired effect*

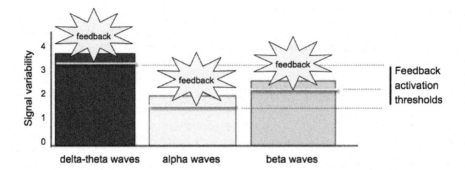

Figure 5.5. *Feedback activation thresholds*
positioned too low. For a color version of this figure,
see www.iste.co.uk/vincent/neurofeedback.zip

This is the case for some hypersensitive people for whom an excess of feedback can provoke discomfort (like a headache, for example) that will wear off fairly quickly. On the other hand, if the thresholds are too high (Figure 5.6), there is no feedback. Then, no training occurs, because the training is directly related to the information given to the brain by the feedback.

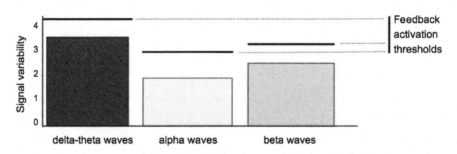

Figure 5.6. *Feedback activation thresholds*
positioned too high. For a color version of this figure,
see www.iste.co.uk/vincent/neurofeedback.zip

5.6. Adapting sessions

It is possible to adjust the tuning frequency manually in order to be as close as possible to the frequency that will be a reference for the others. This is also called the *fundamental frequency*[1].

The tuning frequency makes it possible to regulate the frequencies by "resonance" through harmonics.

Regulating the fundamental frequency (3 Hz) also adjusts its harmonics in 6 Hz, 9 Hz, 12 Hz, 15 Hz, etc. Similarly, by regulating the frequency of 5 Hz, we adjust its harmonics in 10 Hz, 15 Hz, 20 Hz, etc. Note that the frequency of 15 Hz can also be targeted by regulating the 3 Hz and the 5 Hz.

How does this work?

Feedback occurs when a group of frequencies exceeds the activation threshold. The brain cannot understand that this signal is addressed to this particular group of frequencies. On the other hand, it gradually indicates to the practitioner what must be harmonized and the proposed correction made to the brain by the software.

In the previous diagrams (Figures 5.5 and 5.6), the low frequencies (delta–theta) had the greatest variability. Therefore, the feedback sent corresponds to them. When it is triggered, some frequencies are in sync. The brain can then interpret the feedback as a proposal to regulate all frequencies at time "t".

If the feedback is emitted regularly, the brain ignores it because it anticipates it. The period between two pieces of feedback must not be predictable (without being random).

In Figure 5.7, there are three waves of 3 Hz, 6 Hz and 15 Hz. These three waves are in sync three times per second. They are therefore paced by the lowest of them: the 3 Hz. If the feedback occurs when a maximum of frequencies are in sync, this allows for a more extensive regulation. The training is more effective, which explains the search for tuning that corresponds to a maximum of brain waves in sync.

---

1 The fundamental frequency of the note "la" in music is actually 440 Hz and its multiples 220 Hz, 880 Hz and 1760 Hz are its harmonics.

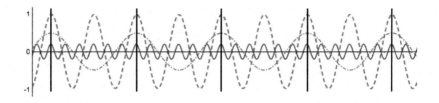

Figure 5.7. *Three waves in the cyclical phase (harmonics). For a color version of this figure, see www.iste.co.uk/vincent/neurofeedback.zip*

The emission of several signals forms waves whose intensities are measured 256 times per second by the neurofeedback system. The following examples show two waves with the same frequency, but different intensities.

Four cases are possible:

– the brain waves are in sync (Figure 5.8). The brain's electrical activity is homogeneous on the frequencies measured. There is no significant variation in the signal intensity. The system is said to be stable;

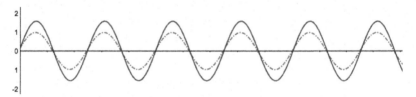

Figure 5.8. *Two waves in sync. For a color version of this figure, see www.iste.co.uk/vincent/neurofeedback.zip*

– the brain waves are in phase opposition (Figure 5.9). The signal perceived is almost zero;

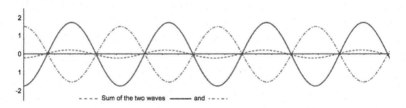

Figure 5.9. *Two waves in phase opposition. For a color version of this figure, see www.iste.co.uk/vincent/neurofeedback.zip*

– the brain waves in sync have too large a range (Figure 5.10): the signals overlap, prompting resonance. The electrical activity is too intense, which can be a sign of a dysfunction: epilepsy, ADHD (attention-deficit hyperactivity disorder);

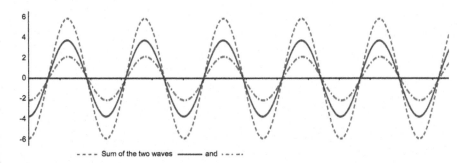

Figure 5.10. *Waves in excess of phase (resonance). For a color version of this figure, see www.iste.co.uk/vincent/neurofeedback.zip*

– the signals are out of phase (Figure 5.11). The pulses are not in sync and the signals received seem anarchic. They are said to be asynchronous. In this case, the signal is difficult to interpret.

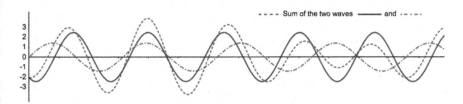

Figure 5.11. *Two out-of-phase waves. For a color version of this figure, see www.iste.co.uk/vincent/neurofeedback.zip*

In a choir, if all of the singers start together in unison, this leads to a harmonious melody.

They can also sing in canon, which involves a delayed start for some of the singers. This delay must occur at a very precise time for it to remain harmonious.

If each choir member sings the same melody at the same rhythm, but each one starts whenever they want to, this creates a sound that is not very harmonious and, in the end, produces a cacophony.

The brain's electrical signals are emitted by neural networks that produce pulses in a nonlinear way, which is to say often out of phase and non-cyclical, unlike tides, whose range depends on the phases of the moon.

The tuning frequency makes it possible to superimpose a reference wave onto the signals emitted by the brain, which makes it possible not to lose the quality of the signal received by the software and to be as close as possible to the signal emitted by the brain. This is like hearing the words of the singers in a choir where no-one is singing in unison. In order to rectify this, the choir master attempts to prompt the singers to harmonize and sing in unison. This is also the role of neurofeedback with brain activity. The tuning frequency is the fundamental frequency of a signal. It was described by *Dennis Gabor*[2].

The Gabor spectrogram, originally defined for light waves, is used by assimilation for the electrical frequencies of the brain by neurofeedback systems. We can compare this adjustment to the one carried out manually on old radios or television sets in order to pick up the correct frequency of a show, eliminate interference and make the sound or image clearer. This adjustment is made automatically on the new radios and television sets.

Another tool that is essential for adapting the session, but does not depend on a specific neurofeedback system, is the medical history. This consists of listing the problems and illnesses encountered by the person who has come for a consultation. This initial questionnaire as well as the summary of what the person felt or perceived between two sessions make it possible to adjust the session so that it is as effective as possible.

---

2 Dennis Gabor: British engineer and physicist of Hungarian origin (1900–1979). A signal consists of an amplitude and a phase, but a normal receiver is not sensitive enough to pick up the phase. In order not to lose this information, Gabor proposed superimposing the signal with a reference wave and recording the interference fringe. The reference wave is called the "fundamental frequency".

5.7. Positioning the electrodes

In order to adapt the session to the patient's situation, we can position the electrodes over different cerebral areas, especially with the *classic neurofeedback* system. This positioning will depend primarily on the aim of the sessions.

It should be noted that the dynamic system requires a standard positioning for the electrodes in C3/C4 (the central left lobe and the central right lobe) above the motor and sensorimotor cortex, while other classic systems require the movement of electrodes according to the symptoms and medical history: this is a symptomatic approach.

A good understanding of the cerebral areas and brain function is necessary for this type of practice.

Here is a universal map of the electrode position points on the scalp as well as their names (Figure 5.12):

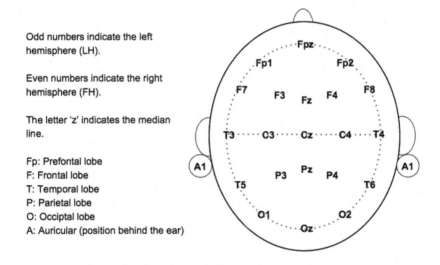

Figure 5.12. *Names and map of points to position electrodes on the scalp*

Although *dynamic neurofeedback* uses two electrodes (and two sensors positioned in A1 and A2 on the ears in order to eliminate the surrounding noise), *classic neurofeedback* uses several electrodes positioned on the scalp according to the aim to achieve. This positioning adapts to the brain function by taking into consideration that several disorders have emotional origins. It is often recommended to regulate the electrical activity of the right hemisphere first – in order to clear out traumas – before regulating the left hemisphere, which manages a large part of the acquisition of knowledge.

The positioning can be modified and can vary over the course of a single session in order to regulate the brain's electrical activity in its entirety, from one cerebral area to the other.

While *dynamic neurofeedback* is based on the assumption that by regulating the central lobe, we regulate the entire brain through resonance, *classic neurofeedback* considers brain activity in several areas, primarily associative ones (see section 6.3).

The origins of illnesses that patients suffer from can be emotional, functional, chemical, structural – or even accidental – but can also have several of these causes at once.

In order to regulate the brain's electrical activity in a gradual way, *classic neurofeedback* proposes positioning the electrodes in order to first regulate the emotions, which are often the primary cause of several disorders. The regulation approach occurs by association areas, from the back of the brain to the front, and from right to left. The brain will therefore be regulated by resonance, as well as by associative areas.

6

Cerebral Areas

The brain is a plastic organ, which means it is perpetually moving and changing, capable of modifying and regulating itself. It is comprised of different parts that interact with each other and each part has a specific function. The human brain differs from the animal brain in that it has an upper layer called the cortex, whose frontal part is very much involved in reasoning, organization, planning and thinking about the future.

This particularity of human beings is what makes us "superior" beings, capable of adapting to our environment quickly and effectively, and one of the only beings to adapt an environment to itself, sometimes to the detriment of other species and even to its own kind, which is perhaps not the best proof of superior intelligence.

Cerebral areas are generally identified by their anatomical construction, which is related to the types of neurons that compose them, or their function, which is their involvement in a specific activity like language, motor function or vision.

Neurofeedback uses these properties of the brain to optimize the results of brain training.

6.1. Dedicated areas

The electrodes positioned on the scalp cannot sense electrical activity further than 3 or 4 cm around them, which implies that their placement on the scalp has an effect on the session's impact.

The brain is structured in cerebral areas and zones each of which has a particular function, while not being independent from one another (Figure 6.1).

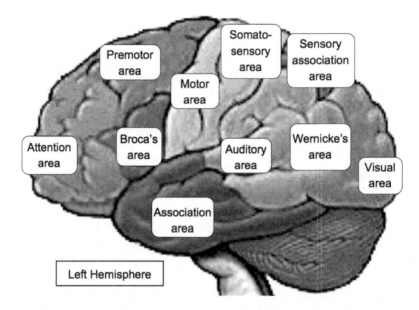

Figure 6.1. *Map of dedicated areas*

We could not imagine a house where all of the rooms have the same purpose and, what's more, could not communicate with each other! Each room has a specific use and is connected with one or more other rooms, although not all rooms are connected with each other.

To put it simply, there are often few doors that open into a bathroom, but a hallway opens into several rooms. Without being essential for its use, its role is crucial for communication.

The same goes for the brain that, aside from the zones called "lobes", has cerebral areas within these same lobes. For example, the parietal lobe includes, among others, the Wernicke's area, which is involved in understanding speech, but also the sensory association area, which adjoins the occipital lobe responsible for vision as well as image comprehension.

"Children, are you ready?"

"Yes, miss."

"I will say a word and I want you to tell me everything it makes you think of. Ready? 'Vase!'"

"Flowers!" ... "The sea!" ... "Soissons!" ... "Pottery!" ... "Fragile!" ... "Old!" ... "My grandma!" (laughter) ... "Vas...eline!" (laughter) ... "Pewter!" ... "Broken!" (more laughter).

"Alright, we're getting off topic! We'll stop there, that's very good! Let's examine your answers."

[...]

"Each of you associated the word with what it makes you think of, by association of ideas. Claire?"

"A vase to put flowers in, like on the dining room table at home."

"What about you, Paul?"

"Me, I said, 'the sea' because I thought of the word 'vast' because the sea is huge. I often go fishing with my dad. I can smell it even now!"

"Yuck!"

"And you, Jane?"

"I thought of history class and of Clovis' famous phrase, 'Remember the Vase of Soissons!'"

"Show-off!" whispered Bertrand in her ear.

And so on, little by little, one idea leads to another. We get off topic, as the teacher said.

These idea associations and even (or especially) the drift off topic that follows are only possible because specific cerebral areas and association areas are connected.

A simple word can trigger a string of reactions in the brain involving sight, speech, memories, writing, sounds, smells and more. This shows that while the specific area triggered was the auditory area, this area immediately transmitted the information to several other cerebral areas through communication paths within the association areas.

The difference in the children's answers depends on each one's life and their relationship to the word proposed. That said, the first response was spontaneous but, gradually, the children sought other associations voluntarily, and enriched the exercise.

This feature of the brain is taken into account in neurofeedback. It is essential to know and understand, even in a simplified way, the role of cerebral areas to better understand a problem and make a session more effective.

> Mr. Y. was having vision problems: blurriness in front of his eyes for a few weeks had forced him to wear glasses. After two neurofeedback sessions with the electrodes positioned at O1/O2 (left occipital lobe and right occipital lobe that correspond to the vision area), the blurriness disappeared and his glasses were no longer needed.

By positioning the electrodes above the auditory area, on the temporal lobe, we have a better chance of getting results for hearing issues and tinnitus than positioning them elsewhere. If the issues are correctly identified, the results can be optimized.

> This is what happened with Ms. C. who had her hearing improve and her tinnitus disappear after seven neurofeedback sessions with electrodes positioned above the auditory area.

However, some issues have complex origins and it is not easy to find the correct positions for the electrodes. It requires experience and attentive listening to the patient because sometimes the problem that led them to you is only a consequence of an older and deeper issue that can be difficult to address to allow the problem that led them there to fade away.

I would like to point out that there is no risk in positioning the electrodes on one area rather than another, knowing that there is an established protocol for different symptoms and positions in *classic neurofeedback* and a progression sequence that proposes enhanced training as the sessions proceed in *dynamic neurofeedback*. It is up to the practitioner to select the parameters that are most suitable for a patient, just like a doctor selects what care is appropriate for the situation of the person who comes to consult them. It is essential that a relationship of trust is established between the patient and the practitioner.

6.2. A symptomatic approach

With *classic neurofeedback*, positioning the electrodes above the associative areas is recommended in order to regulate specific electrical activity in the brain more powerfully because these areas are interconnected with several dedicated areas.

The classic system is unique in that it uses a symptomatic approach to offer a session that is adapted to the symptoms that the patient presents (Figure 6.2).

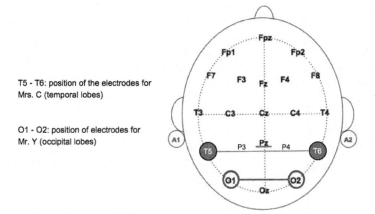

T5 - T6: position of the electrodes for Mrs. C (temporal lobes)

O1 - O2: position of electrodes for Mr. Y (occipital lobes)

Figure 6.2. *Positioning the electrodes for a symptomatic approach. For a color version of this figure, see www.iste.co.uk/vincent/neurofeedback.zip*

On the other hand, the dynamic method is not specially designed for this objective. The position of the electrodes in *dynamic neurofeedback* is C3/C4 in a standardized way.

The symptomatic approach of *classic neurofeedback* requires knowledge about the cerebral lobes and cerebral areas, both dedicated and associative, a consideration of the information provided by the medical history, a good understanding of pathology and, above all, attentive listening to the person's emotions, because each situation, while it may have some similarities with another, is unique and personal.

> With the parents of little A. – with brain damage since birth and aged 15 months at the start of the sessions – we decided on a symptomatic approach concentrating the work on her vision, which was poor, because it seemed to be one of the first manifestations of positive results for her with neurofeedback. I positioned the electrodes at O1/O2, on the left and right occipital areas.
>
> Over time, she seemed more present and better able to follow with her eyes, which the ophthalmologist treating her confirmed with astonishment, unable to explain the sudden improvement.
>
> Little A. had about 40 sessions with considerable progress in areas other than vision as well.
>
> Unfortunately, we did not get to go further in our progress, because she left us a few months after the start of her sessions following pulmonary complications.
>
> "Goodbye, little A. You have a special place in my heart. I will never forget you!"

No matter what profession you practice, it is essential to understand the tools of the trade. It is only with practice and a good knowledge of software and its functionalities that we can optimize the results obtained and adapt sessions to the various situations we encounter.

"I wanted to install an electrical outlet in the wall in my office. It took me a crazy amount of time!"

"But that's not very complicated. What happened?"

"I had to drill a hole in the wall, but it's concrete block. I broke several drill bits."

"Did you have a concrete drill bit?"

"Oh, there are drill bits for concrete? No, I took my drill and I used it as is."

"For concrete block, you need a concrete drill bit. You also need to put your drill in 'percussion' mode, or else it won't be powerful enough, and run at a reduced speed. Otherwise, you risk burning out the motor."

"I didn't know that either."

"And what about your paneling? How did you get through that?"

"With the drill, too. I drilled a hole and used a wood file to make it bigger. It took me six hours!"

"If you had used a hole cutter on the end of your drill, selected for the diameter of the hole you wanted to make, it would have taken a few seconds to make the hole and it would have been clean and neat. My father-in-law always said: 'A good worker has good tools, and knows how to use them!'"

"I'm really a weekend handyman!"

6.3. Association areas

Aside from dedicated areas like Broca's area or Wernicke's area that are associated with verbalization and speech comprehension, or the visual and auditory areas, there are "association" areas. These areas can be considered to be communication paths between different dedicated areas. They are always found between two, or even several, dedicated areas. For example, the sensory association area is found at the intersection between the visual area and the sensorimotor area, among others.

Basically, the individual sees an object, extends his hand to take it and his hand opens to the width of the object, no more and no less, because the brain had a mental image and anticipated the act of seizing it through a process that involved a large number of neural networks. The sensorimotor area is very active as soon as the person makes contact with the object.

The association areas connect the lobes of the brain. The temporal lobe, for example, is connected with the central, occipital, parietal and frontal lobes (Figure 6.3), like the rooms in a house.

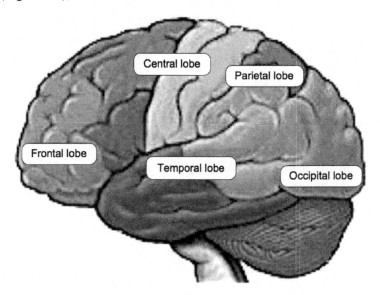

Figure 6.3. *Map of cerebral lobes*

It is thanks to these association areas that neurofeedback can regulate emotions. The cortex, the exterior layer of the brain, includes motor, visual, auditory, analytical and verbalization functionalities, but no area dedicated to emotions. The origin of that is deeper in the brain, at the level of the limbic (or mammalian, common to all mammals) brain. Primary emotions like fear and anger have their own source in the subcortical brain, also known as the reptilian brain, common to reptiles and mammals.

The electrical activity of the mammalian and reptilian brains cannot be directly sensed by the electrodes positioned on the scalp, but these two parts of the brain communicate with the cortex through specific neural connections.

Like we cannot imagine the rooms of a house not being connected to each other, we cannot imagine levels that do not have access to one another.

In the brain, there are connections between the cerebral areas, cerebral zones and the different layers of the brain through neurons, cerebral areas, the corpus callosum, white matter, gray matter, etc. In addition, the brain has two hemispheres, left and right, whose specificities are shared by all human beings and some other mammals. The two hemispheres as well as the frontal, temporal and occipital lobes communicate with each other through a fibrous part called the "corpus callosum" that allows for the transfer of information from one cerebral zone to another and from one hemisphere to the other (Figure 6.4).

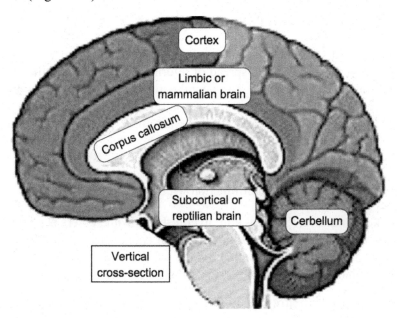

Figure 6.4. *The different layers of the brain*

Damage or poor development of the corpus callosum often leads to speech disorders and motor skills without good lateralization, such as dyspraxia or dyscalculia. It seems that in the absence of serious damage, most issues can be decreased by re-learning basic early childhood skills, like crawling, which requires the corpus callosum to reconnect the left and right hemispheres.

In nature, some animals have an atrophied corpus callosum, such as the kangaroo. Because of this, kangaroos do not have a lateralized gait, but move forward with both feet at the same time.

With neurofeedback, for this type of problem – though I myself have never had a kangaroo attend a session! – we can try positioning the electrodes on C3/C4 (left and right central lobes) above the motor area in order to try to get as close as possible to some specific brain waves: beta SMR (*sensorimotor rhythm*) waves and "high" beta motor function waves.

6.4. The specific nature of brain waves

Information circulates from one area to another, from one zone to another, and the speed of the spread of information depends on the nature of the neurons conveying them. The neural activity translates into the number of electrical pulses produced in one second. One hertz is equivalent to one electrical pulse in one second. If there are three pulses per second, this corresponds to 3 Hz, and if there are forty, 40 Hz, etc.

Therefore, the frequency of the waves depends on the signal emitted, the nature of the neurons and their function within the brain.

Some neurons have a very thin myelin sheath while others have a thicker one and still others have a myelin sheath that is thinner in places. These thin areas in the myelin are called nodes of Ranvier. They promote the acceleration of the spread of the electrical signal in the axon.

The intensity of different brain waves provides information about the dominant activity in the brain. This manifests on the neurofeedback practitioner's screen as an image representing the variations in the intensity of the different frequencies, from 1 Hz to 40–42 Hz, over time.

A node of Ranvier is a thinning of the myelin sheath surrounding an axon in the nervous system. It allows the saltatory conduction of the nerve impulse.

The extension of neurons that propagates the nerve impulse is called an axon. This axon is surrounded by an isolating sheath called a myelin sheath. The points where this sheath is interrupted, leaving the axon "bare", are called "nodes of Ranvier". These are points where most of the sodium (Na^+) and potassium (K^+) channels are concentrated because these areas are where the nerve impulse forms.

The impulse travels from node to node, which considerably increases the propagation speed of the impulse. This mode of displacement is called saltatory conduction.

Box 6.1. *Nodes of Ranvier*

The very high intensity of an electrical frequency in the brain "at rest" indicates to the practitioner that there is major activity on this specific frequency. This can provide an indication about the potential origins of the activity.

It is important that these signals appear at rest, because any movement and speech necessarily involves the activation of a large number of neurons and therefore modifies the information visible on the screen.

During a *dynamic neurofeedback* session especially, it is essential that the person is relaxed and, as much as possible, has their head resting on a headrest, because tension in the neck muscles holding the head upright does not allow the brain to calm down or produce the alpha waves specific to relaxation that support the efficiency of the brain training.

If the person is moving, the electrical activity in the brain is difficult to use to provide appropriate and effective feedback.

We know that blood pressure should ideally be taken sitting or partly reclined, in a state of rest.

Conducting a neurofeedback session with someone who moves and talks is as unproductive as taking their blood pressure in similar circumstances. The measurements do not indicate anything and the feedback has little to do with the actual dysfunctions because a large number of frequencies are very active. It does not allow us to see the peaks of electrical intensity in the brain and correlate the feedback with this electrical activity.

When someone talks and moves during a session, all of the frequencies exceed the thresholds that were initially calculated as the homeostatic thresholds at rest. The software must filter them out, but the markers on the practitioner's screen are sometimes not readable. The process of brain training is not, in my opinion, optimized in such circumstances.

6.5. Contribution of auditory and visual supports

Feedback is correlated to the brain's electrical activity, but only reaches it through sound and image because the neurofeedback software is paired with a multimedia reader, and also with different animation software exclusively present in *classic neurofeedback*.

The feedback can be a disturbance of the sound and the image, a modulation of the sound volume and the clarity of the video, or visual filters positioned on the multimedia reader blurring the image more or less according to the variability of the signal emitted by the brain.

The reward, on the other hand, is indicated by the clarity of the image and sound, as well as by sound and light signals indicating that the response was appropriate, encouraging the brain to reinforce the proposed solution.

However, all of this is only possible if the signal processing software sends the multimedia reading software information that affects its reading parameters. The feedback sent to the brain is therefore, above all, an auditory and visual signal indicating whether the solution is good (reward) or whether it needs to be modified (feedback).

The *classic neurofeedback* software is also accompanied by multiple activities whose visual parameters are related to the brain activity whose threshold exceedances or normal range maintenance are signaled by feedback or visual, tactile or auditory rewards, providing the person with information about their brain activity.

It involves several cognitive exercises, called brain games, that exercise memory, concentration, observation, ability and lateralization, which the patient can carry out during the session. The software then analyzes the brain during difficult tasks and provides information about its activity and progress through auditory or visual feedback.

The tactile reward is a reward system in the form of vibrations felt through a stuffed animal, for example, that the child (or the person) is holding. These vibrations, paired with the correct brain response, give the illusion of purring.

With *dynamic neurofeedback*, it can be interesting to propose music that the client is unfamiliar with during the first session so that they approach the experience from a totally "neutral" standpoint.

Most people believe that music prompts specific reactions or mental images, but the same music can have totally different effects from one person to the next:

> "With this piece of music, I thought I was with the Native Americans."

> "Obviously, this piece of music evokes Africa and the drums."

> "This type of music transports us to the bottom of the ocean."

> "I was dancing with African women who were beating a rhythm on the drums."

> "I saw a group of Natives sitting around a fire. Two of them were standing, beating their drums, with their backs to the rest of the group."

> "I was in Tibet in a village lost in the mountains."

> "It was Siberia, with long stretches of wilderness as far as the eye could see."

And all of this from the same piece of music, listened to by different people!

As a general rule, people do not have a preference for one type of music or another or a certain video support. It is only after a few sessions that the choices become clear and will foster a certain state of mind for them that is conducive for relaxation and letting go and, sometimes, allows access to buried traumas in order to free them or clear them out.

> Mr. P. came in for his second session. I proposed a few pieces of music. He was undecided.
>
> "Why not classical music?"
>
> "Let's try it! [...]"
>
> "This piece of music makes me think of my aunt. She taught me how to play piano. She loved music and from the very first notes, I saw her sitting at the piano. Then, I saw myself at her funeral. It made me want to cry. It was very emotional. I think I haven't yet accepted her passing. This session gave me a sense of liberation. I feel lighter and calmer. So, neurofeedback helps with grieving?"

Mr. P. was trying neurofeedback to attempt to decrease his tinnitus. He never imagined mourning the death of his aunt. And yet, he did. In addition, his tinnitus totally disappeared after a few sessions.

Some of my clients bring their own music that they prefer and with which they have a history of strong emotions.

> That was the case of young C., whose mother brought him to my sessions.
>
> "He has loved this piece of music since he was little," she told me. "I don't know why, but he listens to it on loop!"
>
> "It lasts 3 minutes, so I will play it at the end to conclude the session," I said.
>
> The session unfolded the same as always. Young C. chose to watch a cartoon. His brain activity denoted a certain amount of concentration.

Then came the moment to put on his favorite music. Young C. closed his eyes and almost seemed to be praying. I was surprised, but not as much as when I observed the modifications that appeared in his brain after just a few seconds of music.

"Listen," I said quietly to his mother in order not to disturb her son – although, with headphones covering his ears, he could not hear surrounding noise. "He suddenly has many 3 Hz in his brain, with some alternating from 8 Hz–10 Hz. It's strange, I have never seen this! I'm not sure, but I think your son has put himself into a state of self-hypnosis or something like that. He is no longer here, mentally. It's like he's absent, I think."

"Every time he listens to this piece of music, it's the same thing. He says, 'Mom, this music lets me clear my mind, not think of anything. When I listen to it, everything around me melts away.'"

"Indeed, his brain seems to have completely changed how it functions. It's very surprising!"

At his mother's request, we repeated the experiment and each time, there was the same reaction. However, the mother and I, who were also listening to the music, but through speakers, did not feel anything in particular. It was the score from a movie that was just a normal film, but it had an extremely powerful effect on young C.

It should be noted that when C. listened to this music, the visual and auditory feedback was very pronounced and extremely clear, which has never occurred with anyone else to this day.

I wanted to know more and I proposed this piece of music to several clients of different ages, adults and children. I have never obtained a reaction like C.'s every time he listens to it.

This distinctive effect obviously means something which I am striving to discover. What I take away from this experiment is that the music from this film awakens something powerful in him that he does not know how to define, but that is beneficial for him because notable progress was quickly made after I started to play it for him at the end of our sessions.

However, it is his personal history with the music from this film that plunges him into this state and that history is unique.

With *classic neurofeedback*, the brain training is active and often accompanied by visual activity correlated to biofeedback. The patient is often asked to make a conscious "effort" to regulate their brain, unlike in *dynamic neurofeedback* where the work is done on a subconscious level.

This self-regulation effort can start with an effort to control breathing, becoming aware of issues on a physiological level and voluntarily attempting to modify some deeply rooted habits. This can occur through exercises for memorization, concentration or strategy development, *brain games*, in order to optimize the brain's function. The information provided about the success of these different tests through feedback and reward allows the patient to try to maintain a good state of homeostasis and understand which reaction is the right one and must be stabilized.

The reinforcement loop activated in this way favors the integration of learning. The patient can then try to develop strategies to maintain this state of balance in as stable a way as possible. He then becomes an "actient".

On the advice of a psychologist friend who specializes in EMDR, I experimented with the effects of bilateral music on some of my willing patients. EMDR (*Eye Movement Desensitization and Reprocessing*) is an eye movement technique that makes it possible to desensitize and reprocess information recorded by the brain. This method is very effective for processing recent trauma, for example.

Bilateral music was inspired by this method because it proposes sounds that alternate from right to left, which encourages the unconscious movement of the eyes from right to left, which helps clear out traumas. It is the same movement our eyes make when we sleep, a period that is extremely productive for reprocessing information that reached the brain during the day.

I observed that this type of music had beneficial effects for people: they had more of an impression of having "tidied away" events and having cleared away traumas.

I also examined the effects of bilateral sounds. These are sounds produced at different frequencies that the person listens to through headphones: the sound perceived on the left is different from the one on the right. The difference in frequency between the two will prompt the brain to produce the missing frequency.

There is research being conducted about binary sounds. I am concerned about the relevance of their effects because, while they are definitely beneficial for prompting the brain to produce alpha waves, which are conducive to relaxation, some frequencies, if they are excessive, can induce a state of anxiety. I test these sounds with great care and, in this situation, I am my own test subject.

Part 2

Applications

7

Personal Research

What follows is the result of my observations and empirical research in close correlation with the perception and experience of people who participated in neurofeedback sessions.

Since I started my practice, I have focused on the EEG and the correlation between what I see on the screen – thanks to the breakdown of the signal emitted by the brain waves in a frequency histogram, the different feedback activation thresholds – and what the person feels, as well as what their brain does during the recording.

With this method, I have gathered information about what can occur in a person's brain, without being able to categorically confirm why it is produced in that way. These results are the repetition of events that have been observed time and time again.

Through observations, cross-referencing, hypotheses, deductions, comparisons and questioning patients, I developed a table of my personal interpretation of the involvement of brain waves in various specific processes in the brain (see Appendix 1).

I have observed and noted what I see on the screen, from one person to the next, from one session to the next, since I started my practice and I continue to do so, because there are still so many things to discover about how the brain functions!

I would like to clarify that, in my opinion, the two neurofeedback systems that I use are very effective and that the results described in the following pages were obtained using these two methods, separately or in a complementary way.

Some people only had *classic neurofeedback* sessions while others only had *dynamic neurofeedback* sessions and still others experienced both methods. Patients could also express their preference for either type of neurotherapy.

7.1. The cerebral hemispheres

My observations have led me to verify my hypotheses with some little, sometimes amusing, experiments like the "dancer test" that can easily be found on the Internet.

> The neurofeedback session was over. Mr. C. came to pick up his wife and entered my practice. Mrs. C. had told me that she and her husband hardly ever agreed on anything, and that they had a lot of fights.
>
> I proposed the dancer test to them and asked them to describe what they saw on the screen. Mr. C. told me he could see a curvy dancer who was barely dressed – men! – who was spinning clockwise. Mrs. C., on the other hand, thought that the dancer was spinning counter-clockwise. They discussed it and they argued:
>
> "But look at it! You can clearly see that she's spinning counter-clockwise!"
>
> "But no, she's spinning the other way! Watch my finger, I'm following the movement, you can see it!"
>
> "Me too! Look! I'm following her with my finger and she's spinning the other direction."
>
> "You're talking crazy!"

Mrs. C. turned to me:

"Tell him that I'm right!"

I smiled to see them arguing like this.

"You're both right."

"Impossible!" they cried in unison.

"Yes, because it's a two-dimensional image that is completely black with no depth. It is your brain that creates the depth, and depending on your dominant hemisphere, you will see the dancer spinning in one direction or the other. This is not caused by a dominant eye. The proof is that if you close one eye or the other (if you close both eyes, it's much less successful), you will still see the dancer turning the same direction."

"That's incredible!"

"This could explain why you do not view life or events in the same way. You do not spontaneously use the same functionalities of your brain. For example, Mrs. C., you have an intuitive approach to things, because you are right-brain dominant, while you, Mr. C., have a more analytical perspective, because you are left-brain dominant. From each of your perspectives, you are both right."

"And you? What direction does she spin for you?"

"Clockwise. I am a person who spontaneously uses the left side of my brain. This doesn't mean that I don't use the other hemisphere, of course. But subconsciously – because there's nothing voluntary about it – I use the functionalities of my 'left brain'. But it is possible to work and train yourself to use the other hemisphere consciously. That is one of the possible results of neurofeedback."

This was one of the first observations I made thanks to neurofeedback: the cerebral dominance of a person. I conducted the dancer test (Figure 7.1) with more than one hundred people and the results allowed me to develop statistics that helped them to better understand themselves.

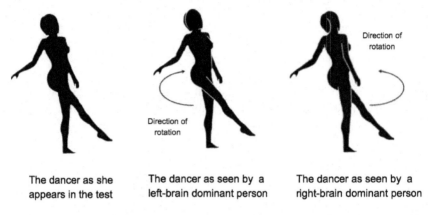

The dancer as she appears in the test

The dancer as seen by a left-brain dominant person

The dancer as seen by a right-brain dominant person

Figure 7.1. *Dancer test and perception depending on cerebral dominance*

Some functionalities of the neurofeedback system make it possible to "visualize" a person's cerebral dominance – if it is significant enough – through the brain's electrical activity. Occasionally, people do not have a specific dominance, right or left; in this case, they see the dancer turning in one direction, then in the other or even balancing without spinning.

It should be noted that, on the Internet, it is said that a right-brain dominant person will see the dancer turning clockwise and a left-brain dominant person will see her spinning counter-clockwise. The experiments that I have conducted have led me to conclude that this statement is false: it is exactly the opposite that occurs!

Without categorizing too broadly, in my observations, it has appeared that the vast majority of right-brain dominant people are women and the vast majority of left-brain dominant people are men.

For children, it is not rare for them to be both, alternately: in this case, they see the dancer spin in one direction first, then in the other.

In a very simplified way, again, men often have a more detailed, analytical and pragmatic perspective of events, whereas women have greater intuition, a more emotional perception of events and a more general view of things. This can, in part, explain existing disagreements between couples, as was the case for Mr. and Mrs. C.

This is not to say that there are no men who are right-brained or women who are left-brained! I myself am left-brain dominant and it is this aspect of my personality that pushed me to want to understand what I was doing with the neurofeedback systems and prompted me to want to decipher what was happening in the brains of people who came to see me, thereby discovering and understanding how my own brain works.

Thanks to this little test, I was able to observe that people changed temperament and behavior over the course of the neurofeedback sessions. This also led to a change in their cerebral dominance:

> Mr. D. had a stroke and suffered from unilateral spatial neglect: he could no longer control the left side of his body. He had always seen the dancer turning clockwise. Yet, at the end of the tenth session, he said to me:
>
> "Am I dreaming? The dancer is spinning the other direction! It's not possible!"
>
> "Try moving your left hand."
>
> He painstakingly opened and closed his fist.
>
> "That's incredible!"
>
> "The next goal is for you to choose the dancer's direction of rotation and to decide on your cerebral dominance."

With this dancer test, I discovered that although we have a natural cerebral dominance, we can also voluntarily use either hemisphere. It demands concentration and self-knowledge that comes with time and training. Then, instead of enduring events, we can choose to rationalize them and draw the strength necessary to move forward.

This capacity of human beings to find in themselves the resources to overcome obstacles in life and survive suffering resembles what the psychiatrist Boris Cyrulnik called "resilience" in his work titled *Un merveilleux malheur* [CYR 12].

In the brain's electrical activity, resilience results in a transfer of the cerebral dominance from the right hemisphere to the left hemisphere.

7.2. Decoding brain waves

Ms. P. had lost her mother six months ago. She was unable to move on.

She came to see me to try to find peace and serenity. Every session was very difficult; she cried a lot, kept seeing her mother. The pain was very intense. From my side, I could see on the screen a majority of 3 Hz as well as a lot of activity between 38 and 42 Hz in her right hemisphere. Her left hemisphere sent no signal of this sort, but signals of frequency association in 3 Hz, 5 Hz and 10 Hz.

I have often had this type of situation with many patients, especially with people in a depressed state. What I observed on the screen corresponded to what people had seen or felt during the session: I asked the person to tell me what happened and, by comparing it with what I was able to observe and by cross-referencing with what others had also told me, I developed a table to decode the dominant frequencies involved in the mental processes (see Appendix 1).

Frequency associations (3 Hz and 38–42 Hz) in the right hemisphere indicate that the person is seeing images from their past.

The activation of three frequencies, 3 Hz, 5 Hz and 10 Hz simultaneously, signals that the person wants to cry. When the person cries, this association is very intense and very dominant in the brain.

Several times, I have said to a person that at just such a moment in the session, I had the impression that a certain sadness had overcome them and I replayed the part of the music during which I observed this frequency association on my screen. It must be said that I was not wrong. It is not divination, as suggested by one client who was a bit too interested in the esoteric, but a correlation between two phenomena that very often occurs from one patient to the next.

> During the eighth session, Ms. P. cried, but felt relieved. At the end of the session, her face was calm and she smiled through her tears.

"My mother came to say goodbye to me!" she told me.

The mourning process had begun.

What had happened on the screen? While Ms. P. was crying and was seeing her mother, her right hemisphere was very active on the low frequencies, associated – in my experience – with buried traumas and unresolved suffering, as well as on the frequencies between 38 and 42 Hz which correspond, from my observations, with mental images and especially memories when the frequencies are associated with 3 Hz.

In the middle of the session, her brain changed how it was operating, moving from 3 Hz on the right to 3 Hz on the left. To me, having observed this event repeatedly, this appeared to be a transfer of information from the "emotional" side of the brain to the "rational" side. For the person, this translates into a sense of calm, liberation and most of all, a feeling that an enormous weight has been lifted off them.

I have often noticed that mourning occurs when the brain transfers intense, emotionally charged information toward the analytical part of the brain, which will rationalize the event, give it meaning and remove the excessive emotional load that prevents the person from feeling well.

> Mr. G. had had a heart attack the previous summer. He came to try neurofeedback out of simple curiosity and did not have any specific expectations. He was quite relaxed and sat down on the chair. He closed his eyes. The session progressed. He opened his eyes again at the end of the session.

"This is an impressive thing! I saw my death!"

"Pardon?" I said, astonished.

I was confused, because at the time, it was the first time that anyone had said that to me.

"I saw my own death! I fell to the ground, and I left my body: I was raised up in the air, I saw myself below, inanimate on the ground, and I felt light and happy as I rose in the sky. Suddenly, I returned to my body, violently, then I left it again in the same conditions, and this back-and-forth happened several times during the session."

I do not claim that I can explain what happened in Mr. G.'s brain. The only thing I can describe is what I saw on the screen as he did his session: many 3 Hz on the right, alternating with 3 Hz on the left, which indicates a transfer of information. Images manifested on the screen because of 38–42 Hz on the right. Most of all, there were many alternating alpha waves in 8 and 9 Hz that indicate, from what I have been able to deduce, intense relaxation close to drowsiness and deep serenity.

Other people have since experienced this type of situation during a neurofeedback session, but of course, in direct correlation with their own life and history: people have seen a near-death experience during a session, although they no longer remembered it. They had experienced heart attacks or had recently come out of a coma.

Others have experienced a phenomenon of disembodiment and mentally floated in the room, seeing themselves from above. I do not know how to explain these phenomena nor the mental processes that cause them. I have only noted them and their recurrence.

It should be noted that these situations represent an exception, in the 0.5% range.

Each session is different and each person may experience (or relive) events that are unique to them.

7.3. Protocol for decoding brain waves

The protocol that I have applied in my investigations on the dominant frequencies in the brain and their involvement in mental processes is the following:

– Throughout the entire duration of the neurofeedback session, I make notes in the client's file about what I observe on the screen:

- the dominant frequencies;

- the frequencies for which feedback is sent;

- the recurrence of the dominant frequencies;

- if the dominant activity decreases or increases during the session;

- if certain frequencies appear or disappear from the dominant activity;

- the alternation of "contradictory" frequencies;

– At the end of the session, the person tells me what they felt or saw, as the case may be, by trying to describe precisely what they were able to experience or identify;

– I compare what they tell me with what I noted down, sometimes asking them to confirm (or deny) my observations and deductions;

– After each session, I take my notes and compare the evolution of the person's brain activity from one session to the next.

I have systematically applied this protocol, which has allowed me to refine my deductions and confirm my observations through comparison from one session to the next and from one person to the next.

I have also submitted this table of the interpretation of the involvement of dominant frequencies in mental processes to about 10 of my colleagues for them to experiment with in turn and make possible modifications or additions to it. This table is part of their work tools for most of them, having been validated by their own observations and ample feedback from their respective clients. This reassures me about the accuracy of my interpretation of this aspect of brain function.

Thanks to this protocol, I have been able to better understand the role of frequencies and their involvement in various mental processes, for which I have not decoded all of the functions. Far from it!

The dominant frequencies are those that have the greatest intensity. They are an indicator of the main activity of the brain at a time "t" and at the place where the electrodes are located. This activity is highly variable, but it often recurs, which means that the brain is constantly managing a very large quantity of information, but its dominant activity indicates what the person is doing consciously or unconsciously.

For example, intense activity between 38 Hz and 42 Hz in the left hemisphere of the person indicates that they are mentally verbalizing or speaking internally. The most dominant cerebral areas activated are the Broca and Wernicke areas, but they are probably not the only active ones.

The feedback is sent to the brain in order to propose a change in an activity or decrease the intensity of the brain's activity. The frequencies to which the feedback "is addressed" are the dominant frequencies, which exceed the homeostatic threshold mentioned earlier and for which the software "estimates" that they are too intense and should be abated. This is the blocking role of neurofeedback.

The recurrence of dominant frequencies is not always an indication of an issue or a dysfunction: it indicates that the brain is focused on an activity that can also become overwhelming and for which the person may have the impression of losing control.

Constant mental verbalization can denote hypervigilance and difficulty relaxing, "emptying one's mind", or letting go. Feedback correlated to frequencies that correspond to this cerebral activity can help the brain to stop its incessant thinking, which can be exhausting.

Another example: the successive recurrence of the frequencies 3 Hz, 5 Hz and 10 Hz, as noted, signals a desire to cry. When this recurrence is too great, the person feels overwhelmed by tears and has the impression that they are no longer in control of themselves.

When the dominant activity increases or decreases over the course of a session, it is a good indication of the brain's reaction to the feedback it has received.

Classic neurofeedback is not only inhibitive. It can also propose that the brain intensifies its activity on frequencies that are deemed to have too low of an activity level (for example, alpha waves so that the person is more relaxed and calm). In this case, the feedback occurs in the zone below the homeostatic threshold and not above it, as is the case for too much intensity.

Conversely, it can suggest decreasing the amplitude of certain waves that are too dominant, again so that the activity is harmonious rather than overwhelming. The 3 Hz is the frequency that all software will attempt to regulate first, because it is often an indicator of "psychological suffering" when it is too recurrent in a person's brain at rest. The feedback will therefore be given for the upper zone of the homeostatic threshold related to this frequency. The aim of neurofeedback is to maintain the brain's electrical activity between the upper and lower limits of the homeostatic threshold.

I have also observed the alternation of so-called "contradictory" frequencies in the brain. I came up with this term because they indicate concepts that are opposite, but revealing, about the individual's personality.

"I have the impression that you are having difficulty getting started in a new undertaking, as if you don't feel comfortable. What do you think?"

"That's exactly how I feel. One minute, I feel confident and I want to try it out, but the next minute, I'm backtracking. You can see that in my brain?"

"Yes, your brain shows an alternation between 40 Hz and 5 Hz, suggesting a desire to get started and an obstacle prompted by a certain apprehension. The 40 Hz is your self-confidence, and it is also Archimedes' exclamation of 'Eureka!' enthusiasm, forward momentum. On the other hand, the 5 Hz is a frequency that indicates a worry or concern or even a certain anxiety, a fear of getting started. It says: 'I wouldn't dare!'"

"That's me exactly: I am always like that!"

"What we can hope is that with the neurofeedback, the dominant 5 Hz will disappear and make room for the 40 Hz more long-term. So that you dare to get started!"

The person may be aware of their brain activity, although not as much in terms of frequencies, as in the analysis of what they can do with their own behavior.

"Today, I feel like I'm at 9 Hz!"

This is how young V., a pharmacy student, expressed herself when I opened the door to my waiting room to welcome her.

"What makes you say that?"

"Since this morning, I feel serene, calm, composed. I am not feeling 'down' as usual. I am relaxed and it feels good!"

I have made the decision to explain to people who come to see me that with my tools, I decode how their brain works (in reality, what is dominant in their cerebral activity, because the electrodes do not sense deeply and we only see a small part of what the brain is doing). I have chosen to make them actors in their neurofeedback sessions. My goal is that they have a better understanding of how they function and of their own history, so that they can better handle what happens to them, and so that ultimately, they become "actients".

8

An Aid for Self-Understanding

We are all shaped by our history and the emotions associated with it.

"Personal history" does not refer to what we remember and what we can recall voluntarily, however.

During neurofeedback sessions, several people have uncovered their very distant past, their early childhood, even their life in utero. This may seem incredible and I must admit that the first time it happened, I didn't believe it.

However, the recurrence of these events with different patients has led me to change my opinion on the subject. After all, I cannot reject a fact that repeats itself simply because I don't know how to explain it. That would not correspond to a scientific approach!

But it is not so easy to push aside the foundations on which we have based our beliefs and personal convictions! I am very "Cartesian", but being "Cartesian" does not mean rejecting what you don't understand. It means seeking out a rational explanation, or at least a plausible and coherent one, for factual findings that can be repeated identically several times and that we therefore cannot ignore.

Mrs. R. experienced something completely exceptional:

"I suddenly had the impression that the music was coming from very far away, as if it was smothered, and I felt like my face was covered in a hot, sticky substance. I seemed to be straining and it was very uncomfortable.

Suddenly, the shroud around me burst, and the sounds seemed to increase tenfold. I nearly cried out!

I have just realized that I relived my birth and that is the source of the hyperacusis that I suffer from daily. I suddenly found myself thrust into a world that was particularly aggressive in terms of sound and that is what I experience every day."

Mrs. R.'s discovery allowed her to understand her problem more fully and tackle it better, making the hostile outside world more manageable for her. The neurofeedback sessions that she attended after this discovery helped her become calmer and eliminate her anxiety related to the constant acoustic assault she suffers.

Mrs. D. wanted to have a child, she yearned for one. She had long sought an explanation for her miscarriages. She had come to do a few sessions out of curiosity because she was sleeping poorly and had anxiety attacks. She did not mention her concerns about motherhood at the outset, because she had not imagined that neurofeedback could help her in that area (neither did I, incidentally!).

During one session, she saw the face of a baby, superimposed over the face of another baby.

"I recognized the face of a fetus," she told me. "But the other child… I don't know."

After a few minutes of silence, she added:

"I think I understand! The other face was that of a baby of one of my friends. He died a few weeks after being born. I have to admit that I am very afraid that the same thing will happen to me. In fact, it already has. Three times."

Mrs. D.'s brain was "blocked" by this event and it was as if she had decided that it was not worth bringing a child into this world to see it die just a few days later. That day, she realized that the stronger her anxiety was, the higher her risk of miscarriage was. She had eight sessions of neurofeedback during which I guided her through a process of respiratory biofeedback with cardiac coherence.

I met her partner two months later. Beaming with pride, he told me that she was pregnant. The baby was born eight and a half months later: "Mother and child are doing well!"

8.1. Stop feeling guilty!

"I've been depressed for years."

"I've really messed up my life."

"I make other people miserable."

"I have dark thoughts."

"It would be better if I killed myself!"

How many times have I heard this kind of talk in my office as a neurofeedback practitioner? Too often! Far too often!

"Trust me. The process will probably take a while, but we'll get there. You will feel better. Quite quickly, I hope. Hang in there! It's worth it!"

What can be said in the face of such distress? I feel powerless.

"Try one session and we'll see how it goes."

"Things can't get much worse".

Mr. M. had not worked in three years. He had lost his job because of the depression he had fallen into. He was "prepared to do anything", but "no longer believed in anything".

The first sessions were difficult. He saw no improvement, and sometimes things were even worse. It took patience and persuasion to keep him from quitting too early and to get him to finish the process.

After seventeen sessions, he felt much better.

But the results only lasted two years, and then he relapsed and came back to see me, feeling guilty that he had not remained healthy for longer.

After eight more sessions, he sent me a message to tell me that he was cancelling our next two appointments. He had gone back to work and he felt great.

I hear from him every once in a while, and his situation has remained stable. He is delighted.

"Good luck, Mr. M.!"

I decided to explain to Mr. M. what I understood from his brain activity so that he could have an indication of the progression of his brain's electrical activity. Unlike the previous examples, this man experienced no images or feelings and therefore had no point of reference to know what his brain was doing or if he was reacting to the feedback he perceived.

"As long as the 3 Hz and 5 Hz are the dominant frequencies in your brain, we must continue. These are what I call 'harmful' frequencies. A 'healthy' brain operates in alpha waves at rest: calm, relaxation, peace, serenity in the dominant frequencies."

Many people (about half of my clientele) are like Mr. M. and do not experience anything in particular during their sessions. It also seemed wise to assist them by indicating the progress that I observed in their brain's activity over time: an increase in this frequency, a decrease in that one, which indicates evolution and modifications that are happening gradually.

This allows them to stop feeling guilty: it is not their fault that they suffer. Their brain presents variations in activity that are too intense and that cause imbalances and disorders for which they are not responsible. They can put what is happening to them into perspective.

By showing Mr. M. his brain activity during a session, his brain became an object external to himself. We were no longer talking about Mr. M. but about his brain, which he had to learn to understand and "tame".

The support that I offer considers the fact that we often ignore what our brain is doing and, therefore, we have no control over how it functions. We sometimes suffer the consequences without being able to act, which can be very frustrating and upsetting.

So many people have stopped feeling guilty of suffering when they find themselves faced with their own brain activity, saying to me:

> "It's not me! It's my brain!"

> "Yes, it's understandable that you are suffering: your brain is too active on the low frequencies related to trauma. Over the course of the sessions, it should gradually learn how to grieve, meaning to transfer information from one hemisphere to the other in your brain. From one room to another in your attic. This is what I call 'tidying the attic'. An organized attic is the start of an appeased brain."

> "You will learn to understand yourself and recognize the signs that your brain sends you when it is suffering, because it governs your actions and how your body functions. This is done little by little, over time, and with patience."

> "Listen to yourself. Trust yourself. I am here to support you."

I call my professional activity *nfba*: *n*euro*f*eed*b*ack *a*ccompaniment (in French: ANFB).

8.2. My brain is fighting back

> "I don't believe in it!"

> "It makes no difference to me."

> "I haven't had any results or improvement in my condition."

> "Neurofeedback doesn't work on me."

Yes, this happens too, alas! Some people do not get any satisfying results from neurofeedback. Or, at least, not the ones they wanted!

Remember that the goal of this method is to help the brain find solutions to problems that are preventing it from functioning well. However, although we have difficulty controlling it, the brain functions in complete correlation with our beliefs and personality.

"Hello, Mr. N. What brings you here?"

"My wife!"

In this kind of situation, I am almost certain that there will not be any major change, because the husband is not convinced of the merits of the method and will have a tendency to prevent things from happening to prove that he doesn't need it. My first instinct is to make him understand that if he is not willing to commit, the sessions will not be as effective and he risks wasting his time and, incidentally, his money!

> "Try to relax, let yourself be carried away by the music. If you need to, try to calm your mind by visualizing some mental images – if you can manage it – related to the music you hear. This may help you to let go."
>
> "I did exactly what you said. I tried not to think too much."
>
> "Indeed, I noticed, but you had to keep repeating to yourself not to think, and in doing so you completely closed off your brain. You did not let go at all, you simply prevented yourself from thinking."
>
> "Was it that obvious?"

Oh yes! I saw an enormous amount of 10 Hz during the session. The 10 Hz is involved in the process of controlling thought in the left hemisphere and emotions in the right hemisphere. In the case of Mr. N., the 10 Hz was dominant in the left hemisphere. He was controlling his thoughts.

There are situations where neurofeedback has few effects and I have directed clients toward other techniques because the results were not satisfactory. Would their brains have resisted so much that it would have taken too many sessions before seeing results? I had issues – possibly unfounded – about advising them to continue when nothing had happened after several sessions because it represents such a financial investment. Neither they nor their close relations really saw any change.

It is important to know your limits and, in this case, the limits of the method.

It should be noted that, currently, neurofeedback is not covered by the social security system in France, although some extended healthcare services do cover a few sessions. In other countries like Canada, the Netherlands and Switzerland, the coverage is much more substantial. In Germany, it is not uncommon for neurofeedback to be prescribed by doctors. This is starting to happen in France.

> A woman called me one day to set up an appointment for her 9-year-old son, who was directed toward neurofeedback by a child psychiatrist. Surprised and thrilled by this approach, I was going to set up the requested meeting when it occurred to me to ask where she was located and who the prescribing doctor was.

> This woman lived in Lorraine and the child psychiatrist was in Germany. I referred them to a colleague a bit closer to her; we were almost 900 km apart!

This experience allowed me to understand that if, in Germany, neurofeedback is already referenced as an effective tool for regulating the brain, its recognition in France is likely to be only a matter of time. At least, this is my hope: the results obtained are very often beyond the expectations of the patients who come for consultation, even if it is obvious that not everything can be resolved with this method.

Some psychiatrists see competition where I, for my part, see complementarity: most patients have already had psychological or psychiatric treatment when they come looking for the help and reassurance of neurofeedback accompaniment because the one complements the other!

8.3. Altered state of consciousness and dissociation

Neurofeedback can contribute to triggering a state of dissociation. Young C., whom I mentioned in a previous chapter, could voluntarily put himself into a hypnotized state, but he is not the only patient with whom this has occurred. What is remarkable in his case is that he put himself in that state all by himself.

In neurofeedback, these kinds of situations happen fairly frequently. They affect about 30% of the people who come for a session and the vast majority of them are women.

The person is present physically. They are not sleeping but, mentally, they are in a second state, as if under hypnosis.

"During the session, I saw a teddy bear. It hurt a lot in my lower abdomen, a pain like a tear."

Ms. D. was crying. She had not yet regained consciousness but she was revisiting an event in her past that was so painful that her brain had blocked it out, erased it to survive despite everything, despite the suffering, shame and disgust.

"A little white teddy bear," she said, "that is crying and has its mouth stitched shut."

I am not an expert in symbols and messages, as I have already noted, but this seemed clear to me. I suggested to her that something very sad and painful may have happened in her childhood that she could not talk about.

"I was raped when I was 8 years old!" she said in one breath. "By my neighbor's grandfather! I have never told anyone."

She told me about her experience in detail, staring into space as if lost in such painful memories. After a few minutes of silence, during which I let her gather her thoughts, I said to her:

"If your brain allowed you to access this memory, in my opinion, it's because it is trying to move past it. Neurofeedback often encourages this. It allows the brain to liberate itself from what is making it suffer unconsciously. It eases the grieving process."

"I have to do something. My life has taken an unexpected turn," she said.

She left, determined to do "something".

Ms. D. came back the following week, smiling and relaxed.

"I absolutely had to find out what happened to my rapist," she said abruptly. "I telephoned his granddaughter, who was my neighbor when I was a child. She told me that her grandfather had died a few years before. I was relieved, I have to say. I told her what had happened. She broke down in tears; she had also suffered the same abuse and had never dared to say anything. We comforted each other and promised to get together."

"Well, you are transformed. Radiant and bright."

"I can tell you now: I have always been afraid of men without understanding why. I turned to women, but that was not the answer either. I now know where my problem started. I was not aware of it."

Fifteen days later, Ms. D. sent me a message announcing her engagement to a man who was madly in love with her and whose advances she had rebuffed for almost two years.

"I wish you all the happiness in the world. You deserve it!"

The phenomena of altered states of consciousness or disassociation make it possible to access buried memories, to "bring them back up" to the surface. Neurofeedback, by facilitating the transfer of information from the right hemisphere to the left hemisphere, also facilitates the grieving process.

There is a strong possibility that when the activity in 3 Hz is very intense in the right hemisphere, the brain is in the process of searching for trauma, whereas when this activity is intense in the left hemisphere, it is in the process of letting go of the trauma. I have observed this type of electrical activity during several sessions and very often the people have confirmed for me that they were revisiting their past and sometimes had the impression that "it was passing by very quickly", like successive flashes, a bombardment of images, faces and events.

In these cases, I imagine a little elf, with only his feet and backside poking out from an immense trunk at the back of an attic, balancing a collection of the most improbable variety of items on his shoulder, moving at top speed. He is obviously looking for something and he does not linger over any of these objects that he might tidy away later. It appears that he has something in mind: he wants to find something specific, but he no longer knows where he put it. So, he rummages... like the brain seems to search through itself during these phases of intense activity in what I call the "attic" of the brain: 3 Hz frequencies.

The process of searching and "tidying the attic" – as I tell the children who come to see me – can take time. That is why I always advise patients to continue as long as the activity on the low frequencies remains intense. As long as the work of tidying is in progress, it would be a shame to stop before completing it because, in a certain way, it consists in a grieving process.

8.4. The grieving process

We might think that grieving for something means accepting the death of a person who was dear to us. This is very often the case but, as we saw above, grieving can also be necessary for rape or abuse. In any case, it consists of letting go of an emotion that is too strong and that is associated with a painful event... which could even be the loss of a blankie for a child!

In this area, the pain belongs only to the person and is unique. It is their story. The intensity of their emotions only belongs to them as well. We cannot judge the impact of an event on their lives for them.

Mrs. B. had suffered from depression for four years, since the death of her daughter, who passed away from cancer. Mrs. B. had the impression that she was responsible because she thought that she did not know how to protect her daughter and save her.

"I abandoned her," she told me.

Why did she experience this death as abandonment?

So, I focused on her history, her childhood and, together, we found the source of her current suffering. Mrs. B. often felt like she was being abandoned or abandoning people in her life. When Mrs. B. was four years old, her brother became seriously ill with a contagious disease.

"My mother left me at my grandmother's house for three weeks while she took care of my brother so that I wouldn't catch the illness," she told me.

As an adult talking, she could rationalize the event but, at the time, as a child, she had only understood one message: "Mom is abandoning me."

It was an act of love and protection from a mother to her daughter that cannot be reproached and yet the child had understood it as an abandonment because she experienced it with the emotions of a child, not the reasoning of an adult.

The feeling of abandonment that she experienced as a child was reactivated with the death of her daughter, triggering a great deal of suffering for her.

The work that Mrs. B. and I did together was effective because of the neurofeedback and because of our discussions after each session, which helped her make connections between the different events that she had experienced. Together, we put the puzzle of her life back together.

8.5. "Everything is set before age six"

As Dr. Dodson said so well, "everything is set before age six." It is also often said that age six is the "age of reason".

What does this mean? By age 6, are children reasonable?

Yes! But "reasonable" in the sense that they can be "reasoned with". In the sense that, at age 6, their cerebral development allows them access to reasoning and understanding words and the meaning of sentences. Before this age, children experience events with their emotions.

Everything is set before age six because the emotions that children experience in their childhood will shape their whole lives. If they do not manage to let go of them and grieve the events that provoked them, strong emotions will remain anchored in them and will certainly be revived when another painful situation arises.

This process of unconsciously anchoring intense events in the brain is what allows memory to function. Without emotion, it is difficult to recollect life events in great detail. The feelings associated with the event facilitate anchoring the memories and remembering them.

A medieval historian once told me this story:

"In the Middle Ages, in the country, a rather cruel way of knowing exactly where the boundary of a land was consisted in placing a child aged four or five on the boundary line, and spoiling them, pampering them, and offering them cakes that they were not at all used to eating.

The child, delighted, radiated with joy and pride to be the center of all this attention.

Suddenly, without any explanation, the child was struck with a huge slap. The humiliation and injustice that the child felt at that moment would permanently anchor the event in their memory.

No matter how much time had passed, they could always say exactly, 'this was where it happened!' and confirm the land boundary in case of disagreements between neighbors."

An undoubtedly effective memory anchoring system, but one that would not be well-regarded nowadays...

Unlike the widely held views that keep people locked in their past, such as "you will never change", "this is who I am", "we are who we are" and so on, it is possible to let go of traumatic events and liberate the brain from emotions and associated issues, especially thanks to neurofeedback.

9

Statistical Information

I have conducted many neurofeedback sessions with very different clinical cases as well as clients of all ages, from very diverse backgrounds.

I noted everything that I saw on the screen as well as what people said to me and told me about how their sessions progressed in order to learn, understand and better help my patients and make progress in my practice.

I thought it would be interesting to compare some similar situations in order to develop a table of interpretation of the brain's electrical activity. At the end of this book, I have included my personal interpretation of the frequency associations in the brain (see Appendix 1). I have tried to translate the language of the brain to better assess the possible effects of neurofeedback on symptoms.

As Dr Elie Hantouche said in the foreword to this book, it is probable that, based on the different frequencies and cerebral areas active during a neurofeedback session, the brain does indeed have its own language and syntax.

Every person is unique and every brain is different, but there are similar dominant electrical activities between two people who have similar symptoms or behaviors. Similarities, of course, but nothing identical.

9.1. Every session is unique

"Hello, Mr. J. How are you feeling since last week?"

"Good, thank you. The session was extraordinary. I felt like I was on cloud nine for the next two days. If possible, I would like to listen to the same piece of music to go through the same emotions and sensations."

"As you wish." […]

"I thought I would experience the same thing as last time. I was ready to be transported through gorgeous landscapes with blazing colors as if on a magic carpet. This time, it was rather gloomy and sad. Are you sure you used the same protocol?"

"I assure you, I set all of the parameters identically: I note everything I do in the client's file each session, all of the parameters that I modify during the session and when I modify them. I also make note of the electrical activity in the person's brain – or the dominant activity where the electrodes are positioned, in any case – and the suggestions that are proposed by the software."

"Then how is this possible?"

At a conference given in 2014, entitled "*On ne se baigne jamais deux fois dans le même sapiens*" (No man ever steps in the same stream of consciousness twice), the neurobiologist Alain Prochiantz said that it is illusory to think that we have the same person before us at any given time. Our cells die, others are created, time passes and our experiences change us [PRO 14].

"During your last session, your brain found and tidied up some events from your life. Today, it did not return to those events."

"But I didn't learn anything about my past this session nor the last one."

"The brain is very complex. From one person to the next, how it works and 'tidies' is different. But – and you just observed this – it can also vary for a single person."

"Do people revisit their past? I only see landscapes."

"About 30% of people that I treat at my practice see events from their past. Others have no 'visual' information. But that doesn't mean that they are not working on grieving for painful events. It simply means that their brain does not 'inform' them about what it is working on. The grieving process takes place without their knowledge, in a certain way."

I became aware of this phenomenon thanks to Mrs. E. She had done eight sessions of neurofeedback and had never seen any images or felt any bodily sensations of any kind, but one day, she said to me:

"I'm done burying my father."

"What do you mean by that?"

"My father died when I was ten years old and since then, at every funeral, I relive his death and its him that I bury. I've been burying him for 30 years. Last week, I went to the funeral of the father of a friend of mine, and for the first time, I noticed that a church could be beautiful. The music was magnificent and for the first time, I listened to it and it soothed me. I felt sad for my friend, but I did not feel the usual pain and I did not see my father's coffin. It was no longer my history that I went through again and again. I am finally liberated and appeased. Yes, that's the word: appeased."

And yet, at each of her sessions, she had not felt anything. The grieving process occurred without her knowledge.

Only the electrical activity in her brain indicated to me that something was happening on the 3 Hz frequency and that her brain was indeed letting go of a painful event.

9.2. Some statistics

My clientele is made up of about 50% women, men 25% and children 25%.

The ages of clients that I treat vary from 15 months old to 89 years old.

These clients are from very diverse social and professional backgrounds: students, unemployed people, business owners, professionals, homemakers, teachers, psychologists, bankers, pharmacists, hairdressers, masseurs, magnetizers, doctors, nurses, farmers, fishermen, cleaning ladies, retirees, artisans, executives, laborers, landscape gardeners, accountants, etc., and of course pupils in elementary, middle and secondary schools, and even children under three years of age.

The reasons why these people come for consultations are also very diverse, although nearly one quarter of them come to attempt to get out of a depressed state.

The most common reasons for a consultation are depression, burnout, stress, anxiety, trouble sleeping, migraines, tinnitus, trouble concentrating, autism, obsessive-compulsive disorder (OCD), Parkinson's disease, Alzheimer's disease, stroke, bulimia, anorexia, hearing loss, learning disabilities, aggression, disorders involving the prefix "dys" (dyslexia, dyspraxia, etc.) and sometimes even marriage problems. Still others come for help to optimize their intellectual or sports performances like concentration, or simply out of curiosity.

Despite these differences and the multiplicity of situations, over the thousands of sessions I have conducted, I noticed that there were "profile types".

Without labeling people or confining them to one category, I wanted to understand why some of them saw images during the session and others did not.

9.3. "Right brain" versus "left brain"

"I was lying in a wheat field, looking at the clouds in the sky, and it was beautiful out," Mrs. O. told me.

"Suddenly, my father appeared, holding me in his arms. I was a baby, and he was feeding me with a bottle. What was he doing in the field with me in his arms? My father was never very loving toward me, I am sure that he never held me in his arms... My father wanted a boy. I was a disappointment."

"I never saw any images during the neurofeedback sessions and yet I know that I grieved for the death of my three-year-old son because I can talk about him without crying, whereas before I could not even mention him without bursting into tears," Mrs. D. confided.

"My mother is transformed. I don't know what you did to her and she herself cannot explain it, but it's the first time I have seen her happy in 30 years. I am 37 years old and my mother has been depressed for 30 years! Since the death of my brothers. It's a miracle!" said Mrs. H. I had come to her house to do home sessions with her mother who, each time, remembered a different death, successively grieving for her brother, her father, and her children.

These situations show that all of these people were able to grieve, but in very different ways and without excessive suffering. Without having to painfully revisit all of these dramatic events.

I always take notes on my observations during sessions, which helps me to build my client files. By cross-referencing, analogy, comparison and deduction, I came to the following conclusion: only people who were right-brain dominant saw images of what their brain was in the process of doing.

Left-brain dominant people, including myself, must be content with the effects of neurofeedback without having well-defined information about what is happening, because they do not often see images or feel bodily sensations.

Our eldest son, who was not often at home and who was not able to have follow-up with neurofeedback, wanted to stop smoking. As a family, my husband, my son and I went to consult a hypnotherapist who was strongly recommended by some of my clients. After all, if you haven't tried it, then you can't talk about it, and you certainly can't judge it!

"A nice set of left-brainers," he said to us as we entered the room.

"How can you tell?"

"I can see it in your eyes! Your gaze reveals the cerebral dominance."

"Oh really?"

"You should know that hypnosis does not work on everyone. Have you ever seen a hypnosis performance?"

"Yes, yes."

"Well, only receptive people go up on stage. In order not to choose volunteers for whom it will not work, the hypnotist – careful, I'm a hypnotherapist, it's different! I heal with hypnosis! – as I was saying, the hypnotist first tests people from the audience. For example, the hypnotist may ask them to cross their hands on their head, then uncross them. The people who are unable to uncross them are receptive and will be able to participate in the performance without risk of failure, which would not be good for the hypnotist's reputation."

"I don't think hypnosis would work on me," said my husband, a professor of mathematics and computer science.

"That's very likely," said the hypnotherapist. "Hypnosis only works on right-brained people and you are left-brained."

"So we came for nothing?" exclaimed our son.

"No, because there is more than one form of hypnosis: there is Ericksonian hypnosis, symbolic hypnosis and mindfulness hypnosis, which does not occur through a hypnotic state but rather through dissociation."

Our son had two sessions of hypnosis and stopped smoking. The most important factor is that he wanted to stop smoking, but that he could not manage to do so. If it had not been self-willed, it would have been more complicated, or even impossible.

I explained to the hypnotherapist what I do and how neurofeedback works. I described how some people "get lost" in their memories and see scenes or people that they had completely forgotten.

"Blocked out," he corrected. "I understand that your system has the same properties as hypnosis and supports access to the unconscious. Like hypnosis. Exactly the same."

"It's possible, but the proportion of people who have access to their memories through images is about one third of my clientele."

"If people are receptive and ready, they will very quickly access the part of their brains which hypnosis starts acting upon to help evacuate trauma or drop addictions."

It was thanks to this hypnotherapist that I started to propose the "dancer test" to people in order to determine whether they were right- or left-brained. This allowed me to carry further my study of neurofeedback.

This does not mean that people who are left-brain dominant are not receptive to neurofeedback, but they should not be surprised when, unlike other people, they do not experience much during the sessions. The effects of neurofeedback on them are more difficult to measure. It is an unconscious change that occurs and the effects are only perceptible to them when they are very obvious.

Sometimes these people, and it is their prerogative, do not attribute these changes to neurofeedback and will seek out a reassuring "rational" explanation for the changes. Indeed, it is difficult to admit that a method that we do not understand could have effects of which we are not conscious.

The main thing is that they feel better, after all, regardless of the reason for their well-being.

It should be noted that if a person is right-brain dominant, this does not necessarily mean that they will see images during the session. The only conclusion that I have come to is that the people who see images are all right-brained, and the opposite is not true.

9.4. "Side effects"

It is important to specify that neurofeedback is *a gentle and non-invasive method – regardless of the system used –* that sends the brain information about how it functions. It never includes acting directly on the brain.

Therefore, a person cannot be harmed with neurofeedback.

The software, through successive feedback, gives the brain a proposition to work with, a suggestion to modify its activity. We can therefore talk about self-regulation by suggestion, although the word "suggestion" tends to make it seem like we are influencing the brain to modify it. In reality, it concerns information that makes it possible to modify a behavior to optimize how the brain functions.

When we have a blood test done, we obtain information about the levels of different markers that are essential for the organism's balance. It is not because the test signals a deficit or an excess that the level will modify itself. However, this information makes it possible to act to get back to normal by modifying diet, sporting activity and, as the case may be, medications.

A second blood test, carried out after a certain amount of time, will provide information about the evolution of the levels of the different markers after action has been taken to regulate them.

If some levels have stabilized, but there is no change to other markers, we can assume that the proposed solutions are good for regulating a part of the dysfunctions observed, but not – or not yet! – for other dysfunctions. We can decide to continue the same protocol or contemplate modifying the solutions implemented.

The same goes for neurofeedback: over the course of the sessions, some issues subside, or even disappear, while others persist and do not seem to change. We can continue with the same protocol, which means keeping the same positioning for the electrodes, the same settings for the filters, and the same thresholds for triggering feedback, or modify them and wait and see what changes this produces. Sometimes, these changes produce unexpected effects that are a sign that the brain is reacting and attempting to regulate itself, without necessarily managing to do so quickly. Nothing prevents the practitioner from modifying the session again in order to get closer to the goal.

With *classic neurofeedback*, it is not uncommon to change the position of the electrodes in order to target first the regulation of the right hemisphere, then the lateralization of the two hemispheres, and finally, only the left hemisphere. The starting premise is that most of the issues come from emotions that are too strong and from grieving that is not over, hence the initial placement of the electrodes on the right hemisphere during the first few sessions.

Although people see their dead parents or scenes from their childhood, it should be noted that these scenes are almost never painful, even if they result in fits of crying. They are often tears of profound emotion and relief.

Other times, the images of the past are very serene and peaceful. I have concluded that the brain, like a computer, "scans" itself to find what it is looking for and will pass over life events at great speed, lingering here or there before continuing its search. It will conduct these searches until it has let go of everything that is disturbing it and that is why depressed people, unlike the majority of other people, often require a greater number of sessions, because their brains have accumulated many painful events that they have not yet managed to grieve over.

To return to the image of the attic, when it is full and we have not taken the time to tidy it, the new elements, as insignificant as they may be, will not be put in the right place and will be stored throughout the house. Depression is, in my view, a house filled with boxes that were never put away and that clutter the space to the point that they obstruct all movement.

The effect of neurofeedback is to help the person to sort everything methodically, but according to tidying criteria that is unique to the person and their brain.

My observations have led me to notice that the way of tidying the brain corresponds exactly to the way people tidy their daily environment.

> Ms. N. had never done any neurofeedback. During her first session, she emptied an entire box of tissues. She cried "all of the tears in her body". She did not know why she was crying. She saw no images. At the end, I asked her:
>
> "When you tidy up, do you put everything on the ground and then sort after?"
>
> "Exactly! How did you know?"
>
> "That's my hypothesis, because your right hemisphere presented such agitation that it seemed like it was 'flinging everything onto the ground'. That's the impression that I had looking at your brain activity on the screen."
>
> "I don't know why I cried, but I feel better."
>
> "Don't be surprised if you have nightmares tonight or tomorrow night: as long as your left hemisphere is not as engaged on the same frequencies as your right hemisphere, it's as if your brain is pulling out a multitude of things, but not putting anything away. Several clients have told me this after a session like yours."
>
> "I hope that next week it will begin to tidy up!"

"That is what we can expect. But it's not an absolute rule! I prefer to let you know so you won't be surprised. At the start of my practice, I did not yet know about all this and I could not predict the potential reactions after a session.

Now, I prefer that people know what to expect, because some were a bit afraid!"

"Oh really? Of what?"

"People are afraid of words. The word 'brain' scares them and they think that we are going to act directly on it and modify it. So, when something unusual happens in their daily life, they tend to attribute it to neurofeedback."

"And that's not the case?"

"It's often the case, but not always."

Although it is not uncommon to have nightmares after a session during which the brain is in the process of searching for trauma, it is not systematic.

The most common phrase I hear from patients is: "I don't know if it's the neurofeedback, but..." followed by a list of phenomena that are surprising for the person, but rarely for the practitioner, because several clients have already noted the same things. At the start of a session, when practitioners ask a patient how they feel and what they have observed or felt since the previous session, we often get this answer: "I don't know if it's the neurofeedback, but...":

"I'm sleeping better."

"I'm calmer."

"I'm not bulimic anymore."

"My appetite has come back."

"I can hear better."

"I'm taking some time for myself and thinking a bit more about making myself happy."

"I dared to tell my boss what's been on my mind."

"I dared to stand up to my husband and not let myself be pushed around."

"I've learned to say 'No!'"

And so on, with this sentence also uttered in an infinite number of ways according to each individual's personality.

"I am not interested in that anymore."

"It's fading away."

"I feel protected by an invisible bubble."

"I'm finally doing what I want!"

"I'm taking some distance."

"I no longer feel hurt."

"I have stopped feeling guilty."

To the point that these changes will affect the person's family and friends.

It is important not to get carried away and think that neurofeedback can fix everything, that any change can be attributed to neurofeedback or that any event is necessarily a consequence of a session: for example, two of my clients broke a limb. One broke an arm the next day, and the other broke a leg three days after a neurofeedback session.

It could always be said that they were troubled to the point of no longer being attentive enough. That's possible. Nothing proves this, but nothing proves the opposite, either, the same way that a party with many drinks, a sleepless night, bad news, an annoyance, an immense joy, or any other event can disturb normal behavior and be the indirect origin of an unfortunate event. Sometimes, it is simply the concurrence of circumstances, like strong rain, slippery ground, etc. that are not due to any human intervention. It's simply a matter of jumping to conclusions…

On the other hand, I believe it is important to signal to people who resort to neurofeedback that it is relatively common to see some "regressive" behavior occur for several days or even several weeks, especially with children, and especially at the start of the process. Indeed, parents bringing in their children hoping for improvements can be frightened by some unexpected behavior that is the opposite of what they had hoped for!

Several parents have told me:

> "If you had not warned us that this could happen, we would have been scared and stopped everything!"

> "Our son has started to have temper tantrums again which he hasn't done since he was four years old!"

> "My daughter has become cranky like when she was little."

> "My son has refused to look me in the eye for a while now."

Or simply, after finishing a session themselves:

> "I had a sleepless night, that hasn't happened in years."

> "I wanted to break everything. I was angrier than ever!"

But also:

> "I've got my little boy back, cuddly and affectionate. It's great! At age 15, we usually get phrases like, 'Mom, you're smothering me, let go of me!'"

> "My daughter often cuddles up in my arms, she's become affectionate again. If only it would last!"

It will probably not last.

What happened?

These are what I call the "side effects" of neurofeedback.

The brain, as we saw, revisits life events very quickly without systematically "sending" images of what it is passing over. What's more, if the client is watching a cartoon or a film – which is often the case for children – a part of their brain is already concentrated on the images in the video. The child will not see other images related to what their brain is trying to let go of.

When the brain finds information that is "not correctly stored" or that is too emotionally charged, it will attempt to put it "in the right place". By reconnecting to this information, it will also awaken the associated emotions. If the event was sad, the person may cry, or if it triggered a lot of anger, the person will be on edge, etc.

We can consider that this entails backtracking on the events to allow new functions to be put in place. This backtracking is often accompanied by regressive behavior associated with events on which the brain is working. This is not negative, but simply troubling for a little while.

The behavior that is produced – which is not systematic – can last anywhere from 24 hours to several weeks depending on the difficulty the brain has in letting go of what troubles it. But we must reassure ourselves: they are always temporary.

They are, in my view, one of the most compelling signs that the brain is working to let go of what is troubling it. If nothing occurs over the course of the sessions, we can legitimately question their effectiveness on that patient.

I believe it is important to highlight that, in light of the cases cited above, it is wrong to believe that with neurofeedback, we simply listen to music or watch a video. It is important to remember that neurofeedback will produce changes in a person and that these changes sometimes induce states of disturbance.

It is essential that this neurotherapy is conducted by a neurofeedback professional. Given the side effects that I have observed with some clients, it is best to have serious and rigorous accompaniment. When the brain is working on letting go of trauma, the person must be able to get an explanation of what is happening.

Some companies are planning to commercialize neurofeedback software for the general public that can be installed on a tablet or mobile phone. Imagine a person walking down the street with their headphones on, but instead of listening to music, they are in the middle of a neurofeedback session! We must hope that none of the undesirable effects that can occur during a session happen to this person!

Although it must be said that it is appealing from a financial point of view, this business idea does not seem ethically acceptable to me from a therapeutic perspective. It would tend to make people believe that the brain can modify itself without ever inducing a state of disturbance, even if it is a temporary one.

I hope to caution potential users against such a practice: one does not improvise the role of the therapist, just like the medical community strongly discourages self-medicating. There can sometimes be significant problems.

9.5. The return of symptoms

Other relatively common manifestations with neurofeedback include the "return of symptoms".

Mrs. F. had been depressed for some time. She had already had 10 neurofeedback sessions and was getting better and better. However, at the eleventh session, she said:

"Neurofeedback is not for me. I'm having dark thoughts again. I'm going to stop! It's nothing against you, rest assured, but it's not for me, it's not working... anymore."

"Since you're here, we'll continue, so that you didn't come for nothing. What do you think?"

"It won't help. I don't believe in it anymore."

"We'll see. It can't hurt… Have a seat, please. I am convinced it will do you good. Trust me."

She accepted. Mrs. F. had a tense look on her face for two-thirds of the session.

On the screen, I again saw many 3 Hz in both hemispheres. Her brain seemed to be once again very active on the infamous low frequencies that are, in my view, a sign of psychological suffering. There had been fewer and fewer in the previous weeks but they were active once again.

In the last third of the session, Mrs. F.'s face became relaxed and calm, and in her brain, the dominant frequencies became alpha frequencies, especially the 9 Hz, a sign of serenity.

"I feel better. That's wonderful."

"It was worth persevering, wasn't it? What happened to you was the return of symptoms. In training, they explained to us that it's as if the symptoms come back one last time to say, 'Goodbye for now.'"

"I would prefer they said, 'Goodbye forever.'"

"That's what I hope for you, with all my heart!"

Mrs. F. attended three more sessions to reassure herself. She had trouble believing it, then stopped her weekly visits: she no longer needed them. She has one "maintenance" session from time to time to verify that everything is going well and to maintain her healthy state. A sort of "technical check", essentially.

"Thank you, Mrs. F., for giving me your trust to the end, in spite of everything."

The return of symptoms is quite challenging, because it plunges the person into a state that they have made an effort to get themselves out of and that more than anything, they fear finding themselves in again. I had not warned Mrs. F. that this could happen because I did not yet know what it entailed. I had not yet had enough patients in this type of situation. Her fear was all the greater because she thought her suffering would continue her entire life! Happily, she has left it behind!

The return of symptoms – when it occurs – takes place generally between 8 and 13 sessions, based on my observations. Sometimes even later, which can be even more surprising.

I believe it is essential to warn people that this can happen at some point in the process.

It is important to understand that neurofeedback is a learning process, a training process for the brain to find or rediscover an ideal and optimal way to function. It assumes that the brain, during the training, discovers a new process or unlearns a bad one. Until the new way of functioning has been adopted, there can be "misadjustments" and backtracking toward what is more usual for the brain.

Between each training session, the learning process continues, with more or less success, because the "teacher" or "trainer" (the neurofeedback system) is not there for guidance. The brain therefore tries what it thinks is the right solution and sometimes reverts back to old, inappropriate processes.

9.6. Suicide attempts

To date, I have consulted with more than 30 people who have made one or more suicide attempts during their lives. I always verify whether the person who arrives with such a history is receiving medical care. I once had to face a very complicated situation that, since then, has strongly urged me to make this verification.

Mrs. K. called me crying. She must absolutely see me urgently, she told me, because she was having morbid thoughts. She begged me.

I found her an appointment as quickly as possible, because she said she could not wait: she was suffering too much. I proposed the following evening at 8 p.m., which seemed too far off to her, but I could not do any better.

The session lasted two hours. I discovered a very complex situation. Mrs. K. needed to talk. A lot. She talked and I listened, trying to understand how she had arrived at this point.

She punctuated most of her sentences with, "In fifteen days, I'm committing suicide, I have calculated everything, planned everything." I know that it is not often the people who talk about it the most who go on to do it and that it is a cry for help, but there was no proof that she would not do it!

I asked if she was seeing a psychiatrist.

"I don't want to see a psychiatrist, and anyway, all of the ones I called are not taking on new patients. They have no more room."

I did not point out the contradiction in her remarks, but I did not want to assume alone responsibility for this distressed woman. It was beyond my abilities.

I directed her toward a psychiatrist, Dr S., who was already seeing some of my patients.

At the next session, she said she had met the psychiatrist and gave her consent for me to contact him to discuss her situation.

It was the psychiatrist who called me. I expressed my concerns faced with the extreme situation of this woman who "cried all the tears in her body" and constantly threatened to commit suicide. I explained to him that I had treated people in very difficult situations before but this seemed too difficult for me and beyond my abilities, and that I did not want to be the only one to treat her.

"You're not alone now, I am also taking her on. Because she has only just started a treatment that I am not familiar with yet, it would be best if she momentarily stopped neurofeedback and stabilized with medication."

"I agree with you, particularly since, like antidepressants, neurofeedback can have a disinhibiting effect after two or three weeks."

"She will likely resume her sessions with you when she is calmer."

"I have seen several depressed people, but what surprises me about this person is that I did not find the 'normal' markers in her brain activity. I wonder what is happening in her brain, because it does not correspond to the 'depressed types' that I have interacted with up until now."

"You are right to note this. I don't know what indicators you have in neurofeedback, but I can confirm that this person is suffering from an atypical depression, which could explain why you did not see the normal markers."

"Thank you for your time, Dr S. I have to admit that it is rare for a doctor to take an interest in neurofeedback, and even rarer that a doctor takes the time to contact me even though most of them know that I am seeing their patient in sessions."

"It comes down to respect, quite simply."

Respect. I believe that this woman, mistreated by life, had a great need for it and that she was lucky to come across this psychiatrist who was respectful of her pain and her person, and who responded to me with so much humanity.

If Dr S. reads this book one day, I hope he recognizes himself!

"Thank you for your respect, Dr S."

9.7. The number of sessions

During the first session, most people ask for an estimate of the number of sessions it will take them to achieve their aim. It is almost impossible to predict the exact number needed to obtain the expected results. The effects of neurofeedback depend on several factors.

First of all, the desire to change. This may seem absurd because we might imagine that if people have taken this approach, they want to change. But it happens that this entails so many changes to a well-established routine – despite how painful it may be – that it disturbs them too much and sometimes pushes them to stop the therapy. Change can be scary.

Some clients come, urged by others who told them how much the method helped them, and expect to benefit from the same results. Yet, while they are sometimes similar from one person to the next, the results are very closely tied to the individual and their life.

Moreover, the plasticity of the brain that neurofeedback draws on varies from one person to the next (see section 5.2).

In addition, in order for the results to be quickly felt and lasting, it is essential to attend sessions regularly, once or twice a week, at least for the first few weeks during which the modification process is put in place. Because it is a learning and training process, if the sessions are too spaced out, we lose many of the benefits of the training. The same is true for maintaining physical fitness.

It happens fairly often that people come in for particular issues and see other symptoms disappear:

> Mr. and Mrs. G. both attended a few sessions. He came because he had tinnitus and hearing loss, and she came because she was anxious and wasn't sleeping well. They were both over 80 years old.

> After six sessions, Mrs. G. told me:

> "I was so scared! I thought my husband had died!"

"What happened?"

Mr. G. was there in front of me. I had to wonder. He seemed in good health for someone who had had a brush with death!

"For three nights now, my husband has stopped snoring! The first time, I was so frightened! I thought he had died. But as I got closer, I heard his breathing, and that reassured me!"

This is an example of one of the "unexpected" results of neurofeedback. The brain reorganizes itself according to its own priorities that we don't have control over.

Mr. G.'s tinnitus decreased after eight therapy sessions and his hearing also improved.

The number of sessions necessary to reduce tinnitus depends entirely on the type of tinnitus the patient has and their brain's plasticity.

Mrs. C. came to see me because she had tinnitus and hearing loss. She was "trying her luck", to use her own expression.

From her first experience, she was astonished:

"I thought that the music was getting louder throughout the session. I could hear better and better!"

During the second session, she told me she had to have her hearing aids serviced because they were no longer suitable. The hearing test showed clear improvement in her hearing.

At her third session, she almost no longer needed her hearing aids.

"I can hear well enough. The hearing aid specialists don't understand it."

I'm afraid they are not doing any advertising for neurofeedback. They risk losing a few clients.

However, the case of Mrs. C. is rather exceptional. Few people have their hearing improve so quickly, although the phenomenon is relatively common after several weeks of neurofeedback. However, it would be misleading to say that everyone who comes to see me for this type of problem have their issues disappear. Unfortunately, it is not that simple.

> Mr. F. had had tinnitus for several years. He attended about a dozen sessions, without any notable improvement in his symptoms.

> "I feel calmer, I am much less stressed so my tinnitus doesn't consume me anymore and it's no longer my top concern. I live with it very well, now."

He did not reach the goal he had come in for, but the results obtained allowed him to live with his problem better.

10

Clinical Cases

Some of my colleagues have told me that they quite often have clients come in for personal comfort or even to optimize their cognitive and intellectual performance. It is true that more and more high-level athletes regularly practice brain training with neurofeedback in order to increase concentration, reduce stress and ultimately be better than their competitors. This is not very well known, because if the information spread, other athletes would also begin such training and the advantage would be lost.

Therefore, I will not divulge here the names of athletes using this method which has the significant advantage of having almost the same effects as doping (the comparison is probably a bit excessive), without inducing negative effects on their health as well as their professional careers.

I have only had a few cases of people resorting to brain training for mere pleasure, although most of them come to like it.

The majority of people who come to consult me are suffering a great deal, even to despair, such as patients in depressive states that are sometimes very serious.

10.1. Neurofeedback and depression

Because I have observed it many times, I can confirm that it takes between 20 and 30 sessions at least to get out of a depressed state, with some rare exceptions.

I have many depressed patients and the majority of them also receive psychiatric care. Some, who are hospitalized, come to consult me during the hours they are allowed to leave the hospital, sometimes on the advice of their psychiatrist.

About 15% of these people have already made one or several attempts to commit suicide before consulting me.

About 90% of these people recover from their depression in an average of 25 to 30 sessions, 2% in under 20 sessions and 3% had to have more than 40 sessions. Another 2% progressed well for a long time and relapsed inexplicably, but have continued to attend sessions, recuperating little by little the benefits provided by neurofeedback.

On the other hand, 3% of the people did not go sufficiently far in the process and abandoned it after four or five sessions. The results at this stage were obviously not effective enough on the depressed state. The people who have been encouraged by parents or friends who have already experienced the method for themselves are more determined, because they are surrounded by people for whom the results have been very satisfactory. This explains why the vast majority of my clientele is made up of people who learned about the existence of neurofeedback by word of mouth.

One of my clients came "forced" by his family and did not believe in the method, or perhaps expected results too quickly. This impatience, or his skepticism, caused him to stop much too early, which unfortunately did not provide any significant results.

I present these statistics to my depressed patients, hoping that they too will obtain fast results that meet their expectations.

It must be added that some studies have been conducted about neurofeedback and depression that encourage the use of this method in these situations:

> "Many patients show no or incomplete responses to current pharmacological or psychological therapies for depression. Here we explored the feasibility of a new brain self-regulation technique that integrates psychological and neurobiological approaches through neurofeedback with functional magnetic

resonance imaging (fMRI). In a proof-of-concept study, eight patients with depression learned to upregulate brain areas involved in the generation of positive emotions (such as the ventrolateral prefrontal cortex (VLPFC) and insula) during four neurofeedback sessions. Their clinical symptoms, as assessed with the 17-item Hamilton Rating Scale for Depression (HRSD), improved significantly. A control group that underwent a training procedure with the same cognitive strategies but without neurofeedback did not improve clinically. Randomised blinded clinical trials are now needed to exclude possible placebo effects and to determine whether fMRI-based neurofeedback might become a useful adjunct to current therapies for depression." [LIN 12]

It would be foolish to believe that all disorders will disappear after three or four sessions. The brain does what it can with what it understands of the feedback that is sent to it and especially with the load of suffering from which it must free itself.

I would like to thank the increasing number of doctors and psychiatrists who trust me and trust the technique of neurofeedback for directing some of their patients toward the method to support medical treatments.

In particular, I would like to thank those who took the plunge and came to test this neurotherapy for themselves in order to be able to better discuss it with their patients.

It should be noted that, since 2016, there has been a national neurofeedback day in Paris (one of the partners of which is the French psychiatry convention) that is attended by many therapists and psychiatrists.

I recently had the agreeable surprise of noting that neurofeedback appealed to some of them for their personal well-being. This open-mindedness can only contribute to greater progress in the effort to improve health conditions for people who are suffering psychologically.

I would like to point out, nevertheless, that 2 to 3% of the people who have come to see me did not achieve the expected results or even any conclusive changes, despite the fact that they had come for help with the same issues for which other people obtained very good results. To date, I have not found a satisfying explanation for this. It is, in my view, a limitation of the method. As the hypnotherapist said:

"Some people are receptive, others less so."

10.2. Lasting results

People have told me:

"One of my friends said that she had attended some neurofeedback sessions, but that the results didn't hold out."

"What do you mean by that?"

"She had five sessions and was sleeping better, but then, things went back to how they were before. She said that it doesn't work in the long term."

"That doesn't surprise me much. The results that can be obtained must be anchored to be lasting. In five or six sessions, we can make some changes, but they will not be stable: this is training for the brain, which must learn a new way to function and how to keep it up so that the changes stabilize. Your friend probably needed to attend more sessions more regularly. How many was she having per week?"

"She had one every now and then, when she could."

"I always advise that sessions be attended regularly and not be too spaced out from one another so that results are quick and lasting. I believe this is essential."

Indeed, it is common for people to stop as soon as they feel better and that is very natural. However, it is important not to forget that this is a habit to form; that the routes taken by the brain to self-regulate are new connections and that they must be taken as often as possible to reinforce them so that they become the new way of functioning.

Have you ever noticed how a path forms in a field?

When the field is free from any track, it is the world of everything possible, it is the child's brain where everything has to be built. Every path is possible. The child can learn every existing language. If the child is immersed in a French-speaking environment, their little brain will specialize in the sounds of this language to the detriment of other languages. It will take a path that will become THE path to take.

When we have to cross a field, we instinctively take the route that is already there, because it has been traced and the grass is trampled, which leads us to believe that the path leads somewhere! We generally don't venture into the tall grass without knowing where we are going.

Growing up, learning languages becomes more complicated because it entails taking new paths that do not or no longer exist. This leads the brain to new experiences and new experiments.

Experiments and experiences often involve trial and error. We throw ourselves into the long grass a bit haphazardly and we see where it leads us. Sometimes, we come upon a pond and we must go around it, or find too dense of a thicket and we must retrace our path.

The brain does the same thing and tries new connections, new routes. If the path taken is the right one, it will continue along this path and, by dint of taking it, this path will become THE path to take.

But as long as the other path exists, there is a risk of making an error and taking it by mistake. The brain can fall back into how it used to function if it has not sufficiently tested the new route and memorized the new path. It must become a habit and, above all, the old route, which is no longer appropriate, must be gradually erased.

For this to happen, we must wait for the grass to grow back without anyone walking over it and trampling it anew.

Neurofeedback is a guide for the brain. It is a signal that indicates to the brain: "Not that way!" each time that it tests a direction that is not correct. With each piece of feedback, the brain receives this information and attempts to take it into consideration by modifying how it functions, but it can also ignore it and make its own decision and go as far as this experiment leads. This will take more time. The classic reward system, on the other hand, indicates whether this is the correct path and whether the brain should persevere in that direction. Every brain finds the most suitable path according to how it is organized and how it functions.

Having conducted many neurofeedback sessions, I have made the following observations:

– the plasticity of the brain is intimately linked to the person's capacity to adapt to new things and manage the unknown;

– changes are produced more quickly if the person is capable of changing their point of view, correcting mistakes, recognizing when they are wrong (for example) and especially, accepting what they do not always understand.

Mrs. H. brought in her son M. who had behavioral issues.

"We can hope that after a few sessions your son will be better behaved in class and less aggressive. I have seen several children with similar issues and for the most part it has worked well."

"Oh, I know him, he'll never change. He has always been like this. His father is the same!"

"Like father, like son?"

"Exactly."

"So, why propose neurofeedback to him?"

"You never know…"

"Perhaps you should encourage him at home."

"Oh, but I do that already!"

"Explain the situation at school, ask for him to have more time, because we are not going to solve the problem in two weeks."

"I've already done that, but no one listens to me. The teacher said that it was not up to her to make the effort."

Needless to say, young M. did not experience many substantial changes. He had already been put in a box: "He'll never change", "Neither will the teacher..." In addition, his problem was considered "hereditary". Still, he did more than 20 sessions... for nothing, or almost nothing.

Conversely, little N., age 5, whom the specialists said would probably never be able to read or write, was brought in by his mother who believed in him. He had brain damage from birth and the medical prognosis was not optimistic.

"Do the sessions last very long?" asked his mother. "Because he can't sit still for more than five minutes."

"I will start increasing the duration of the sessions a little bit at a time."

"I'm tired of this!" exclaimed N., tearing off the headphones and the electrodes. Things were flying everywhere. Obviously, ten minutes was still too much for him, even though he was watching a cartoon that he had chosen himself.

"We will try to get him to hold on for longer next week."

The next three weeks were not as successful as anticipated and I was about to lose hope of obtaining results. But one day...

"Is it over already?" little N. asked me, surprised, at the end of the tenth session (no less!).

His session had lasted 30 minutes. As long as he was not moving, I let the time tick by and prolonged the brain training. It could only help!

He started having sessions in June. In September, he started elementary school accompanied by an educational assistant to help him work, because it was difficult for him to hold a pencil, write and especially read. He also continued his neurofeedback sessions.

At the end of November, his mother told me that he no longer needed the educational assistant and almost knew how to read and write. He attended 17 sessions at a rate of two per week at the start, then one per week for a while and finished with one every fortnight.

He came back to have a session that I call a "technical check" and everything was going very well. He has not needed to be seen again. Everything was back in order and yet, he had *a priori* fewer chances than young M. of working through it.

For N., the results were fairly spectacular and, above all, lasting, which is not always the case, it must be said.

To achieve long-term effects, it is best to have weekly sessions and not to hesitate to come back for a few more if the symptoms do return, even if only mildly. It is preferable not to let them settle back in to avoid having to come back too often.

I recommend a short session every once in a while to check that everything is going well, especially when a patient still feels fragile and the results have not yet fully stabilized. This is the so-called "technical check" which has above all a comforting and reassuring role.

A healthy brain will not show any sign of disturbance on the frequencies that I call "harmful": the low frequencies. A healthy brain is dominant in alpha waves when at rest with the eyes closed. The dominant electrical activity is between 6 and 9 Hz. Calm, peaceful, relaxed and serene.

However, it is important not to lose sight of certain pathologies that manifest through precisely such an excess of alpha waves in the brain. This is the case of epilepsy. Alpha waves should be dominant in a healthy brain, but not invade it. Too much of anything can be harmful.

10.3. Neurofeedback and epilepsy

If there is one pathology on which the inhibitive effect of neurofeedback is appropriate, it is epilepsy.

Neurofeedback is all the more effective when the brain is calm and the person is relaxed. Yet, it is precisely in this state that epileptic seizures are at risk of being triggered. A person who suffers from epilepsy cannot easily relax and settle down. This can trigger a seizure if they are not stabilized by medication.

Epilepsy, as I understand it, is characterized by an imbalance between two neurotransmitters: gamma-butyric acid (GABA) and glutamic acid (glutamate), and it manifests as an excess of alpha waves in sync. The software detects all of the anomalies in the brain's activity and sends it feedback each time the harmony is disrupted, each time there is an excess repeated on a given frequency.

During an epileptic seizure, GABA (which is responsible for the relaxation response) is in excess, while the glutamate (an excitatory neurotransmitter) is in deficit: this is exactly the scenario for an imbalance in the electrical activity. The software sends feedback until the brain calms its activity on these frequencies.

Sessions can trigger a seizure to better "fight" it. This can be distressing, because an epileptic seizure can be frightening (especially for the practitioner), but it is important to know that, in the long term, the seizures should be less numerous and less violent, until they gradually disappear.

If the person is taking stabilizing medication, there is little risk that they will have a seizure during the session despite the relaxation in which they are gradually immersed. I myself have never had to manage this type of situation. A study of people with epilepsy showed rather encouraging results:

> "Beneficial effects of NF training on seizure occurrence have
> been described in epileptic patients." [OST 10]

It should be noted that many people see neurofeedback sessions as a time to relax that calms them and they may only come to relax, or even sometimes to optimize their cognitive performance. It is a special time during which these people take care of themselves, like Mrs. O.:

"I do not have *a priori* any specific symptoms. I feel very well, but this is my 'me' time. I relax, I let go and it does me so much good that I never miss a session. Some people go jogging, others do yoga. For me, it's neurofeedback that brings me the most peace!"

10.4. Neurofeedback and Alzheimer's and Parkinson's diseases

I have had a few cases of epileptic patients, five at the most, and even fewer people with Alzheimer's and Parkinson's diseases. Four to date, to be exact.

I would still like to address the topic, because there have been very interesting, if modest, improvements.

There has been better presence in conversation, improved spatio-temporal references and better short-term memory for Alzheimer's in the space of 12 to 16 sessions.

It should be noted that the working memory also gradually returned for people with brain damage due to head trauma.

Young G., age 30, came out of a coma following a very serious traffic accident. Two years after this terrible event, he came for a consultation. Every session, I had to explain again what would happen. He kept forgetting. His memory did not seem to extend beyond 24 to 48 hours.

"You will have headphones over your ears, you will be immersed in a visual environment on your screen and..."

"I know, there's no need to explain that to me," he interrupted rather abruptly. "But I don't want the same environment as last time, I didn't like it," he added in a reproachful tone.

"Last time? That was fifteen days ago! And you remember?"

"Yes, and?"

"Normally, you always ask me what will happen at the start of the session, what this is for, why this, why that. Your memory is improving! And so are your motor skills! I noticed that it took you many fewer steps to get from your wheelchair into the consulting chair."

"Hey, you're right!" he said, and his face lit up. I had never seen him smile before.

I also observed the return of a certain motor agility and a slight improvement in diction after 9 to 12 sessions in the two cases of Parkinson's disease.

A remarkable and unexpected result occurred with one of the Parkinson's patients: he no longer has diabetes! This patient had been diabetic for several years and was being treated for it. His doctor discontinued his medication, noting the return to normal in his blood workup.

One of my colleagues reported the same results for three diabetic people she was treating with neurofeedback.

These improvements can be explained by the fact that neurofeedback is a tool for regulating brain activity. Through self-regulation, the production of neurotransmitters and neurohormones balances out, triggering a return to a homeostatic equilibrium for the entire organism. Anything is possible, even if these cases may be the exception. Time will tell!

10.5. A diagnostic tool?

Neurofeedback is not a diagnostic tool. Its purpose is not to identify a given pathology. For me, it is an indicator of what is going on in the brain, like a thermometer indicates if there is a fever, without providing the reason. The brain, presented with a reflection of its own ways of operating, will try to modify itself, in the same way that with a bad fever, an organism will fight to lower its temperature, occasionally with the need for outside intervention.

However, just as a battery of tests can give clues about the possible issues and make it possible to refine a diagnostic, the different dominant frequencies in the brain can give indications about the state of the brain at a time "t", which can point to potential changes, by comparison and analogy, with a fairly low margin for error but with some uncertainty nonetheless, it must be said.

I do not use neurofeedback to diagnose anything, but to help people to better understand what is happening in their brain, especially when they experience neither feelings nor images, which is the case for two-thirds of my clientele.

However, I don't believe it is out of line to state hypotheses about the origins of the disturbances in the electrical activity appearing on the screen.

During my training in neurofeedback, I noticed a peak of 36 Hz in the brain of my colleague with whom I was testing the system. Looking at the screen that she was viewing, I observed a tree that had part of the trunk missing, giving the impression that it was floating in the air.

At the end of the session, I asked if something had surprised or disturbed her.

"Yes, at a certain point I saw a tree without a trunk and I was startled. Why do you ask? Did it show on the screen?"

"Yes, I saw a peak in 36 Hz that reveals, in my view, concern or surprise. That's why I was hoping you could confirm it for me."

Similarly, a lack of control in the brain can be discovered by attentively observing the markers present on the screen (like the 10 Hz, especially), but this can in no way be considered to be a diagnosis for a given pathology, although certain substances seem to influence the lack of activity on this frequency (alcohol and cannabis, notably). Nevertheless, trying to help the person find better control seems to be a worthwhile goal.

This goal can be worked on in several ways: with neurofeedback of course, and also with a psychologist if there is one, as well as family, friends, etc.

Young J., age 9, had temper tantrums. She lost all control and became violent and abusive. She could not stand her little brother's teasing. She thought it was deeply unfair that he should be allowed to bother her all the time.

The sessions calmed her, but did not prevent the temper tantrums from occurring from time to time.

I tried to work with her on what she thought was unfair.

"What does your little brother do that seems so unfair to you?"

"At the dinner table, he gets up and taps everyone on the back, so I yell and insult him."

"He taps everyone and so everyone yells and insults him?"

"No, only I yell."

"How do you explain that only you react this way if he is clearly bothering everyone?"

"Because the others don't care!"

"But you? You care?"

"Yes, it annoys me!"

"We're going to play a little game: imagine that deep down inside you, there is a little elf (or a little bug, if you like) that pushes you to get angry, because it likes that. It is happy when you get annoyed because you get scolded and that makes it laugh!"

"A bug? A black bug, then!"

"Alright, a black bug. So, each time you want to yell and get angry, try to stop yourself so you don't please the black bug and observe how the people around you react to your little brother. It might give you an idea about how to react differently."

The next week, the mother told me that J. had changed her behavior. When her brother bothered her, she spoke to herself, saying:

"No, no, no! I will not give you this pleasure, I will not get angry." And she laughed instead of getting annoyed!

"And how did her little brother react?"

"He stopped bothering her. He used to enjoy seeing her get angry... and get scolded!"

Sometimes, discussion can resolve situations that seem to have stalled. I resort to it when it seems relevant, although I am not a trained psychologist or psychotherapist. I simply use what could be called "common sense".

Conversely, what appeals to some clients, men in particular, is precisely that neurofeedback is not psychotherapy and that it is not necessary to talk:

"If I had had to talk, I never would have come."

Clients have said this!

I have heard this short sentence from the mouths of three clients for whom about a dozen sessions resolved the psychological problems that had brought them in without me ever knowing what preoccupied them. The medical history did not indicate anything. According to them, everything was going well: they had absolutely no desire to confide or talk. "Secret defense!" in some way.

No problem! Neurofeedback works just as well without any talking.

Thank goodness!

10.6. Neurofeedback and multi-handicap

I have had the opportunity to work with young people who are multi-handicapped from 80 to 100%. Young people who cannot express themselves, like infants.

We might imagine that the results on this type of subject would be minor, the brain being in most cases damaged and impaired. And yet, changes have occurred.

Some people believe neurofeedback only has a placebo effect. So be it. Why not, after all? Doesn't the doctor have a placebo effect as well? Most of the time, merely being given medical care helps people to feel better. I would even go further: merely booking an appointment allows people to feel better, because looking forward to care helps to improve the person's state, and all the better! The placebo effect is beneficial! "It's not a bad thing if it helps."

But can we talk about the placebo effect on multi-handicapped people who do not have an awareness of what is happening? And yet there were changes. Not huge changes that would radically modify the lives of these young people, but all the same, improvements that facilitated their everyday lives and those of their families.

> K. is a young man who is 17 years old now. He was 15 when I experimented with neurofeedback with him. I say "experimented" because it was the first time that I had treated a multi-handicapped person. His parents told me that they expected nothing, but were ready for anything.

> The modifications that occurred with him were remarkable, although they may seem insignificant for anyone who is not familiar with multi-handicap.

> "He has stopped inflicting harm on himself, grinding his teeth, drooling constantly, rocking from right to left, and he has become more responsive, less tense, and sometimes meets your gaze; he seems more present," his mother tallied after about 20 sessions.

Nothing too spectacular if we were expecting him to recover his motor function or speech, it's true, but still improvements that prove that even a damaged brain can modify itself and find new connections to move forward.

This opens encouraging perspectives in which the placebo effect has no place.

10.7. Neurofeedback and ADHD

Studies conducted on neurofeedback and its effects on ADHD (attention-deficit hyperactivity disorder) are quite numerous and show rather satisfactory results, although often with a large number of sessions.

> "Neurofeedback participants made more prompt and greater improvements in ADHD symptoms, which were sustained at the 6-month follow-up, than did CT participants or those in the control group. This finding suggests that neurofeedback is a promising attention training treatment for children with ADHD." [STE 14]

I have treated about 10 children diagnosed with ADHD. The results obtained varied greatly and not all of the children presented the same cerebral characteristics – far from it.

Some saw their symptoms disappear completely after 16 to 20 sessions of neurofeedback, others did not show any conclusive changes. I have a few hypotheses regarding this.

The children suffering from ADHD whose diagnostic was relevant presented similar ways of functioning, namely large electrical pulses in the brain for a duration of two to three seconds at most, followed by a relative return to "normal" functioning. "Normal" means, in this case, normal for these children. The general electrical activity of a brain presenting the issues of this disorder is excessive on almost all brain waves: the amplitude of the variation of the frequencies is very great and often exceeds $10\ \mu V$. It would seem that the entire neuronal system is unstable and disturbed.

Yet children diagnosed with ADHD who did not present these dysfunctions obtained results fairly quickly. I have come to the conclusion, which may be wrong, that there was an error in the diagnosis. My experience has shown me that it takes time to treat these kinds of symptoms.

Although this does not relate to neurofeedback, I also noticed that these so-called "hyperactive" children often had an unbalanced diet almost exclusively based on products containing sugar (cakes, cookies, pastries, soda, fruit juice, etc.).

In some cases, diet can slow or even prevent the results of neurofeedback on the brain's self-regulation.

Moreover, for some troubles where people did not notice a specific improvement and where sugar did not seem to be the only cause, I have suggested that families experiment with a gluten-free or lactose-free diet for several months. The results did not take long to show!

10.8. Neurofeedback and sleep disorders

This is a challenging issue to address.

Challenging, because sleep disorders are often multifactorial: they can be caused by an emotional disorder (such as difficulty letting go of traumas, even small ones), physiological dysfunctions (digestion and breathing, for example), an imbalance in the brain's electrical activity (hypervigilance), a dietary excess or imbalance, conflicting medications, even "simply" the side effects of some treatments, or poor health practices – and sometimes, all of these at once.

Regarding hypervigilance, inquiring about the medical history in the first session as well as visualizing my patient's brain activity on the screen have allowed me to make the following observation: most of the time, hypervigilance affects women with one or more children.

In my view, this can be explained by the fact that by bringing a child into this world, the mother becomes responsible for it and, because of this, her brain starts up a monitoring process, like a radar, allowing her to wake up when her child cries. This is something that many fathers are reproached for not doing, although for the most part, they are not equipped with this surveillance radar, and they really do not hear the child cry. Of course, some of them also pretend not to hear the child so they don't have to get up in the middle of the night.

However, hypervigilance has also been discovered with several of my male clients.

These men all have something in common: they all have a high-risk job or one where they are likely to be woken up at all hours (the military, police, firefighter, doctor, etc.). But also, one totally independent category from this list: the "doting fathers" who, after the birth of their child, develop the same radar function as the mothers.

The results of neurofeedback on sleep disorders are not very consistent and few studies have been conducted on the topic so far:

> "A significant improvement in Total Sleep Time (TST) was found only after the neurofeedback (NFB) protocol. Furthermore, sleep logs at home showed an overall improvement only in the neurofeedback group, whereas the sleep logs in the lab remained the same pre to post training. Only NFB training resulted in an increase in TST. The mixed results concerning perception of sleep might be related to methodological issues." [COR 10]

Sleep disorders are not always easy to treat, especially if one of the causes is dietary.

Once again, I noticed that people for whom neurofeedback had no effect on sleeping troubles often had a dietary intolerance or even allergy. Most of the time, gluten seemed to be the cause. This hypothesis was verified by removing all products containing gluten (comprised of two proteins: gliadin and glutenin) from the diet for at least one month and then reintroducing the protein. After removing gluten, sleep was deeper and more restorative and after it was reintroduced, the trouble sleeping came back.

For my own part, I removed gluten from my diet as a last resort: since then, I no longer have trouble sleeping (I had suffered from insomnia since the birth of my first child, which tends to verify the hypothesis about the origin of hypervigilance noted above). I also no longer have joint pain related to ankylosing spondylitis, nor abdominal pain.

For me, neurofeedback had the effect of greatly reducing my stress level and lowering my blood pressure, which has so far remained normal: I have not required treatment of any kind for several years.

When my sleep is disturbed again, which rarely occurs, one neurofeedback session is generally enough to get everything back in order. It should be noted that I had about 30 sessions before achieving these results, which seem to be lasting.

I have suggested this habit of attending a neurofeedback session from time to time to many of my clients who are inclined to come as soon as they start to feel a persistent instability in the quality of their sleep again, or the reappearance of any other symptom.

However, there are symptoms that are very difficult to eliminate or even to reduce.

10.9. Neurofeedback and autism

I have treated children and adolescents with varying levels of autism. Some children who could not speak, some who could express themselves in a very broken and laconic way, others who could express themselves quite well with mostly natural but relatively mechanical speech, and finally, some with Asperger's syndrome.

I believe this is an important distinction to make, because the expected, or rather desired, changes are very different depending on the level of autism.

I do not have enough distance from the possible results with autism, but my experience with these young people has taught me a great deal about the potential effects.

The results shared by most of these young people were progress with communication, both verbal and visual, and an appeasement that made tantrums due to anger and irritation less common, even if sometimes, alas, some tantrums were not prevented, causing considerable damage in some families.

The road is long in such situations, because how the brain normally functions is deeply anchored in ritual, a difficult element to modify. Changes are perceived as destabilizing and disturbing, and the brain very quickly returns to what it knows how to do, to its codified and almost immutable habits. But changes do occur, even if they take much longer to take effect. Studies conducted up to this point on neurofeedback and autism do not make it possible to demonstrate the effectiveness of this method, although empirically the results are very real, even if slow to settle in, on the few people that I have received for consultation.

I would now like to turn to one of my Asperger's syndrome "patients" who left a moving testimonial on my website about what neurofeedback has given him. Here is what he wrote:

> "Neurofeedback helped me a lot. I was very stressed, I suffered from events that traumatized me (abandonment by my father, humiliation at school). But thanks to neurofeedback, over many sessions, this suffering is now only a vague and distant memory. There's no denying that it is effective, even if it can appear 'magical'." (Samuel, age 23)

This testimonial moves me because it was spontaneous and delivered "as is". Most of all, it touches me because it comes from my very first neurofeedback "patient" for whose well-being I trained in this therapy: my son.

11

Testimonials

Samuel was the first to attest to the changes he obtained through neurofeedback, but he was not the only one. Here is what some of my other patients have said:

"I found an attentive listener as well as solace, followed by a clear improvement in my depressed state. I recommend this method." (L., age 59)

"I resorted to neurofeedback three years ago. In a few sessions, I noticed very surprising results. I gained perspective about events, the ability to let go, better self-confidence and self-esteem. I rediscovered my positive attitude!

Thank you, Pascale, for your professionalism and your knowledge, your active listening and your kindness." (A., age 46)

"A thousand thanks for the invaluable neurofeedback sessions that allowed me to relax deeply, calm my troubled state of mind and better center myself.

I noticed the benefits of the sessions very quickly, both personally and professionally. They allowed me to increase my ability to concentrate and my mental performance." (V., age 47)

"A year ago, I attended a few sessions of neurofeedback with Pascale Vincent.

Despite being a bit hesitant about this method at the start, I have been convinced by the benefits that I gained from it: relaxation, clearer ideas, better sleep, re-centering myself. I have only obtained positive things from it." (F., age 69)

"Taking time for myself, feeling myself grow as a person, strengthening myself, and calming the turmoil of thoughts and emotions to allow life to move freely in all of my cells. This is a short list of what neurofeedback gave me. It provides me with much more and each session opens up a new dimension, but it is also Pascale Vincent's support, her ability to listen and her rich and compassionate way of being there that bring about these unique moments.

She introduced me to the incredible tool for well-being that neurofeedback is. I immediately wanted to test it out of curiosity. I tried it and, like chocolate, I adopted it." (N., age 44)

"Three years ago, I had ten neurofeedback sessions because I was very anxious. I had insomnia and sometimes felt uneasy. After the sessions, I felt much more relaxed and my sleep improved. I greatly appreciated the sessions, especially because they allowed me to let go of difficult events from my past without making me relive them." (E., age 63)

"Thanks to the neurofeedback that I had two years ago, I am able to sleep well again after twenty years of suffering from insomnia. No therapy had worked up to that point. I am therefore very satisfied with this method of care and I am finally free from trouble sleeping." (V., age 51)

"After about ten neurofeedback sessions, my migraines and tinnitus started to disappear. I noticed an improvement in my ability to concentrate and I had less trouble going to sleep. A huge thank you to Pascale for proposing this therapy to me, because the doctors that I had consulted before had not found any physical anomalies. Very often, everything resides in the brain. I advise everyone to try it because, in every possible way, it can *only* have a positive effect." (Y., age 25)

"Neurofeedback gives me so much serenity that I recommend it to all my friends here and in Paris. Over the course of the sessions, feelings, sensations and memories resurfaced in my memory, they were so deeply buried that only when they returned to my consciousness could I sort through them and let them go, which provided great relief. The brain sorts and the body feels better: improved sleep, no more nightmares, smiles returned and laughter too... Any crippling guilt has disappeared.

The goal of my sessions: to rediscover the joy of living. Achieved!" (A., age 72)

"Neurofeedback is an effective method of care that brought me inner peace. I feel less aggressive and less nervous. I can more easily manage my anxiety and express my emotions. I react better when faced with unexpected situations.

I can contemplate future events more calmly. I am more positive in my thoughts." (C., age 48)

"What is better than sitting comfortably in a chair, listening to music that Pascale has chosen for us or that we have chosen with her and letting go or even sleeping and then, an hour later, feeling invigorated and free from emotions that are too strong, and rediscovering serenity, responsiveness, good spirits, enthusiasm? This is what I found in the neurofeedback sessions, thanks to Pascale, her attentive listening and her availability. Thank you so much." (J., age 50)

"I had a few sessions of neurofeedback about two years ago. I felt calmer. Now, I have a session every now and then when I feel stressed because of my work, to help me clear my head." (P., age 49)

Conclusion

In conclusion, to recall the image of the orchestra, neurofeedback is the conductor who listens and indicates corrections to be made in the musicians' performance in relation to the sheet music placed in front of them, recorded and transcribed by the software.

Dynamic neurofeedback is a conductor who listens to all of the musicians together, and indicates when an error should be corrected. Each musician must understand who the signal is addressed to through self-analysis. This is a global and passive approach: the client does not do anything during the session.

Classic neurofeedback is a conductor who, in addition to the possibility of listening to all of the musicians together – like *dynamic neurofeedback* –, also has the ability to address instruments by family so that they gradually regulate themselves. This approach is symptomatic and active: patients are actors in their therapy.

Neurofeedback is a computer tool that promotes the regulation of the brain and, as a result, the regulation of the entire organism. When the neural activity is consistent, the production of neurotransmitters balances out, which tends to gradually stabilize the organism as a whole and causes many symptoms to disappear.

Most people who have attended enough sessions for the alterations to be effective and lasting have attested to the radical changes in their lives. Most of them have resumed normal professional activity and now have thriving personal and social lives.

This method also changed my life: I have practically no tinnitus any more, I am more calm and relaxed. I sleep very well after 20 years of insomnia and, above all, I found the energy to change professions. Thanks to neurofeedback, I have become...

... a practitioner-clinician of neurofeedback!

Appendices

Appendix 1

Frequency Table

A1.1. Personal interpretation of the implication of dominant frequencies in certain mental processes in human beings (empirical research)

LH = left hemisphere; RH = right hemisphere.

Hz	LH/RH	Implication	Examples
2 Hz	LH/RH	Physical discomfort, inflammation Physical pain	Itches, need to urinate, etc. Back pain, inflammation, etc.
3 Hz	LH/RH RH LH	Irritation Looking for traumas Letting go of traumas	Disagreeable sounds, etc. Death, separation, abuse, etc. Liberation from associated emotional stress
5 Hz	LH/RH	Concern, anxiety, anguish	...
6 Hz	LH/RH	Relaxation	...
7 Hz	LH/RH	Schumann resonance (?)	
8 Hz	LH/RH	Mental relaxation	Drowsiness, falling asleep, etc.
9 Hz	LH/RH	Serenity	Sensation of floating, weightlessness, etc.
10 Hz	LH/RH LH LH RH	Control Recurring idea Controlling thoughts Controlling emotions	Controlling thoughts, emotions, etc. Repetition of an idea, phrases, etc. Preventing oneself from thinking, etc. Preventing an emotion from "rising", etc.
11 Hz	LH/RH	Concentration Being anchored in the present	Perception of sound, listening, etc. Thinking about the present, etc.
12 Hz	LH/RH	Auditory concentration	Attentively listening to sounds
13–16 Hz Beta SMR	LH/RH LH RH	Physical relaxation Physical relaxation on right side Physical relaxation on left side	Relaxing the shoulders, the back, etc. If the person is right-handed If the person is right-handed

14 Hz	LH/RH	Schumann resonance (?)	
18 Hz	LH/RH	Joy, happiness	
21 Hz	LH/RH	Schumann resonance (?)	
23 Hz	LH/RH	Obsessions	Intrusive or recurring thoughts, etc.
24 Hz	LH/RH	Motor skills	Body movements
25 Hz	LH/RH	Motor skills	Body movements
26 Hz	LH/RH	Motor skills	Body movements
27 Hz	LH/RH	Motor skills	Body movements
28 Hz	LH/RH	Schumann resonance (?)	
35 Hz	LH/RH	Schumann resonance (?)	
36–42 Hz	LH	Thinking in words, verbalization	Talking in one's head, etc.
	RH	Moving images	Film, evocation, imagination, etc.
36 Hz	LH	Verbalized interrogative mode	Asking oneself a question, making a hypothesis, etc.
	RH	Projecting into the future	Visualizing scenes to come, imagining, etc.
37 Hz	LH	Thinking in words, verbalization	Talking in one's head
	RH	Moving images	Film in one's head
38 Hz	LH	Verbalized declarative mode	Talking in one's head, commenting, etc.
	RH	Moving images	Film in one's head
39 Hz	LH	Thinking in words, verbalization	Talking in one's head, etc.
	RH	Moving images	Film, evocation, imagination, etc.
40 Hz	LH/RH	Self-esteem	Having confidence, motivating oneself, etc.
	LH	Exclamatory mode	"I know", "I've got it", "Eureka!", etc.
	RH	Moving images	Visualizing oneself succeeding, etc.
41 Hz	LH	Thinking in words, verbalization	Talking in one's head, commenting, etc.
	RH	Moving images	Film in one's head
42 Hz	LH	Thinking in words, verbalization	Talking in one's head, commenting, etc.
	RH	Moving images	Film in one's head

© ANFB Accompagnement NeuroFeedBack

COMMENT.– The content of this table is the result of my observations, deductions and personal analyses.

This work of researching and interpreting frequencies is independent from the training I received in the Zengar and Othmer methods.

A1.2. Decoding frequency associations

LH = left hemisphere; RH = right hemisphere.

– 3 Hz + 5 Hz (LH/RH) = anger;

– 5 Hz + 10 Hz (LH/RH) = worry, fear;

– 3 Hz + 5 Hz + 10 Hz (LH/RH) = desire to cry (fleeting if not very frequent, tears if frequent);

– 8 Hz (RH) + 11 Hz (LH) = alternating between drowsy/alert (difficulty going to sleep);

– 6 Hz + 8 Hz + 9 Hz = serenity, "zen attitude";

– 3 Hz + 10 Hz (RH) = psychological block, a trauma does not "let itself be found". This is an unconscious phenomenon;

– 3 Hz + 40 Hz (RH) = self-esteem block;

– 3 Hz + 10 Hz + 23 Hz (LH) = intrusive psychological block. Conscious phenomenon;

– 3 Hz + 38–39 Hz (RH) = images of the past;

– 3 Hz + 5 Hz + 38–39 Hz (RH) = "painful" images of the past;

– 11 Hz + 37–38 Hz (RH) = images of the present;

– 37–39 Hz (LH) + 38–39 Hz (RH) = commentary on images;

– 10 Hz + 23 Hz = "intrusive" obsession (if too frequent, a tendency toward obsessive compulsive disorder);

– 5 Hz + 10 Hz + 23 Hz = "intrusive" and distressing obsession;

– 6–8 Hz (LH/RH) + 38–39 Hz (RH) = dream (if there are no other dominant frequencies);

– 13–15 Hz (LH/RH) = relaxed muscles;

– 6–8 Hz (LH/RH) + 13–15 Hz (LH/RH) = mind and body relaxation;

– 10 Hz + 40 Hz (LH) = verbal motivation, encouragement.

Nota bene

– If 3 Hz is very frequent in the RH: probable depressed state.

– If 3 Hz is very frequent in the LH: the brain is working on evacuating traumas (which neurofeedback is directing the brain toward).

COMMENT.– Some people *only think in words*: verbalization of thoughts (many 38–42 Hz in the LH and very few 38–42 Hz in the RH). Left-brain dominance.

Some people *only think in images*: little verbalization (many 38–42 Hz in the RH and very few 38–42 Hz in the LH). Right-brain dominance.

Some people think *in words* at certain moments, and *in images* at other moments (phases of 38–42 Hz in the LH and 38–42 Hz in the RH). Alternation between left/right-brain dominance.

Some people comment on the images that they "see" (alternating between 38 and 42 Hz in the LH/RH). Left/right balance.

Appendix 2

My Career Path

What is important to me is to introduce and spread knowledge about neurofeedback and its benefits while not hiding its limitations.

I was a teacher in the national education system for 18 years and then I resigned from my post to become a neurofeedback practitioner.

After receiving training in the NeurOptimal® neurofeedback method, I opened my consultation practice.

Always keen to learn and better understand my activity, I received training about the brain, how to read electroencephalograms, and about disorders for which neurofeedback can be a significant help.

Then, I received training in the Neurofeedback/EEG-Biofeedback Cygnet® system to gain a symptomatic approach in line with my own observations and experiments.

Currently, I work with both systems, which are very complementary and which achieve very good results for many disorders.

NeurOptimal®, the *dynamic neurofeedback* system in which auditory feedback is very effective, is, in my view, more appropriate than the classic system for young children, multi-handicapped people and visually impaired people. I have found it to be very effective for regulating depression disorders, among others.

Cygnet®, the *classic neurofeedback* system in which visual feedback is very effective, is, in my view, more suitable for disability issues, like disorders that start with "dys" or ADHD, for example. In addition, it instantly informs the person about their brain's electrical activity thanks to the reward system that encourages the patient's active participation in their therapeutic care. I have found it to be very effective especially for issues related to memory, motor skills and concentration.

During sessions with my patients, I use the tool that I believe to be the most appropriate for the situation of the person who has come to consult me, unless this person has a preference for a particular system.

Moreover, in order to optimize my practice and the support that I can give to my patients, I have received training in several techniques that are different from neurofeedback and that I believe are complementary for a holistic approach to the person: Touch For Health®, Knap points and trigger points.

I also pursue my investigations and expand my knowledge about the brain and neurofeedback at the *Centre de formations et de recherches internationales Neuroptimum*® (Neuroptimum® International Research and Training Center) where I have received training related to psychopathologies, the brain and its diseases and the support relationship.

I also intend to seek training to use the LoRETA (Low-Resolution Electrical Tomography) tool and to obtain the BCIA (Biofeedback Certification International Alliance), a certification that is internationally recognized by the AAPB (Association of Applied Psychophysiology and Biofeedback), the BFE (Biofeedback Federation of Europe) and the ISNR (International Society for Neurofeedback and Research).

Appendix 3

Acknowledgments

I would like to thank Catherine Caillon, one of the pioneers in the open practice of neurofeedback in Brittany and in France, who introduced me to this method that has allowed me to set up as a practitioner-clinician of neurofeedback.

I would like to thank Daniel Wagner, Zengar representative in France and trainer of *dynamic neurofeedback* according to the NeurOptimal® method (originally from Canada), for the high-quality training that he gave me in this neurofeedback system and his humanist approach to the method.

Daniel Wagner is a neurotherapist and founder of the *Centre de formations et de recherches internationales Neuroptimum*® (Neuroptimum® International Research and Training Center) in France.

I would like to thank Philippe Gauffriau, a Neurofeedback/EEG-Biofeedback trainer of the Othmer method (from Germany), for the quality of the training about the brain and the symptomatic approach to neurofeedback with the Cygnet® system, EEG Info, that he provided me with.

Philippe Gauffriau is an occupational therapist and clinician of Neurofeedback/EEG-Biofeedback in Mönchengladbach (Germany).

I would like to thank my son, Yoann Vincent, a sound technician, for his professional contribution to signal processing and the use of specific vocabulary.

I would like to thank my husband, Pascal Vincent, associate professor of mathematics and professor of computer science, for his help in developing the graphs and his clear explanations about the laws of mathematics that govern signal processing.

Appendix 4

Dr Elie Hantouche

Dr Elie Hantouche is a psychiatrist who has specialized for more than 20 years in bipolar and obsessive–compulsive disorders (OCD). He runs the *Centre des troubles anxieux et de l'humeur* (Center for Anxiety and Mood Disorders) in Paris (www.ctah.eu).

He is the secretary for the *Forum bipolaire européen* (European Bipolarity Forum) with Professor Jules Angst and Professor Giulio Perugi (www.eubf.org), an organizing member of the IRBD world conference, the annual meeting of international experts on bipolar disorders (www.irbd.org), and a scientific advisor to AFTOC (aftoc.club.fr) and the information site www.bipolaire-info.org.

He has hosted more than 800 conferences and organized more than 20 international symposiums.

He is the author of more than 200 publications, 61 of which are indexed in PubMed and eight of which are about OCD, anxiety disorders and bipolar disorders, including:

– *Troubles bipolaires, obsessions et compulsions*, Odile Jacob, Paris, 2006;

– *Cyclothymie: Troubles bipolaires des enfants et adolescents au quotidien*, Josette Lyon, Paris, 2007;

– *La cyclothymie, pour le pire et pour le meilleur*, Robert Laffont, Paris, 2008;

– *L'Anxiété*, with the CTAH collective, Josette Lyon, Paris, 2008;

– *Troubles bipolaires: manie, hypomanie et dépression*, Meditext, Paris, 2008;

– *Soigner sa cyclothymie*, with V. Trybou, Odile Jacob, Paris, 2009;

– the trilogy *Méthodes anti-phobies*, with V. Trybou, Josette Lyon, Paris, 2009;

– a new edition of the book *Comment vivre avec une personne atteinte de TOC*, Josette Lyon, Paris, 2005, entitled *TOC, vivre avec et s'en libérer*, with C. Demonfaucon and V. Trybou, Josette Lyon, Paris, 2009;

– *Les Troubles bipolaires pour les nuls*, First, Paris, 2017.

Two of his areas of interest include improving the screening and management of OCD and bipolar (cyclothymic) disorders and, above all, facilitating access to information intended for doctors and the general public.

Glossary

AAPB: *Association of Applied Psychophysiology and Biofeedback.*

Activation threshold: a limit above or below which an alert signal called feedback is triggered.

ADHD: Attention Deficit and Hyperactivity Disorder.

Alzheimer's disease: Alzheimer's disease is a neurodegenerative disorder of the brain tissue that leads to a gradual and irreversible loss of mental functions, in particular memory. It is the most common cause of dementia in humans. It was described by the German doctor A. Alzheimer in 1906.

Asperger's syndrome: Asperger's syndrome is part of the spectrum of autistic disorders, but without the delay in the appearance of language or intellectual deficit. This syndrome was named after the research of the Austrian doctor H. Asperger in 1944.

Autism: autism, and more generally the disorders on the autistic spectrum, are Pervasive Developmental Disorders (PDD), characterized by limited social interaction and communication, and stereotyped and repetitive behavior.

Axon: the axon connects the cell nucleus of the neuron to the synaptic zones (see *synapse*). It is composed of myelin (see *myelin*) and channels (see *channels*).

Biofeedback: feedback about biochemical measurements, heartbeat, skin impedance, body temperature, blood pressure, breathing rate, etc.

Channels: sodium and potassium channels are where ions are exchanged in the myelin membrane that makes up the protective sheath around the axons. Their opening determines the polarization of the neural system and the activation of the impulse.

Dopamine: dopamine plays a role in motivation by acting as a reward system. This molecule is also involved in some pleasures, such as listening to music.

Electrode: a sensor positioned on the scalp that is able to sense electrical impulses emitted by neurons.

EMDR: *Eye Movement Desensitization and Reprocessing* is a technique using eye movements that reprograms information stored in the brain in order to let go of traumas related to post-traumatic stress.

Feedback: a return of information.

Hormone: a substance secreted by the organism and released into the bloodstream. It acts in a specific way on one or more target organs to modify how they function.

Hypervigilance: a state in which the brain receives too much information from its environment.

Ion: an ion is an electrically charged chemical element that has lost or gained one or more electrons. An ion is not electrically neutral.

Medical history: a questionnaire that indicates the different problems that could be encountered by a person seeking treatment and that will make it possible to implement a management protocol.

Melatonin: melatonin is a neurohormone synthesized by serotonin (see *serotonin*) and involved in regulating the sleep cycle.

Multi-handicap: multi-handicap is an accumulation of disabilities and impairments of diverse origins: severe motor and mental disability, for example. The consequence of multi-handicap is an inability to live autonomously and express oneself or communicate independently.

Myelin: myelin is a substance that is primarily composed of lipids. Myelin isolates and protects nerve fibers.

Neural activity: all of the electrical exchanges that are produced between neurons.

Neurofeedback: a return of information about the brain's electrical activity.

Neuron: an excitable cell that makes up the cerebral system and transmits electrical signals in the nervous system.

Neurotherapy: methods of care that focus on the activity of the neural system (see *neuron*).

Neurotransmitter: neurotransmitters, or neuromediators, are chemical compounds that are released by neurons and that act on other neurons, called post-synaptic neurons.

OCD: Obsessive–Compulsive Disorder.

Parkinson's disease: Parkinson's disease, as described by J. Parkinson in 1817, is a degenerative neurological disease (gradual death of neurons) affecting the central nervous system that is responsible for slower movements, tremors, rigidity and cognitive disorders. It is the second most common neurodegenerative disorder after Alzheimer's disease.

Reward: a reward system.

Serotonin: a neurotransmitter notably involved in regulating psychiatric disorders like stress, anxiety, phobias, depression, etc.

SMR: Sensorimotor Rhythm. It consists of brain frequencies related to motor and sensory activity.

Stroke: cerebrovascular accident.

Synapse: a synapse is a functional zone of contact between two neurons, or between a neuron and another cell, that converts information transmitted by the pre-synaptic neuron into a signal in the post-synaptic cell conveyed by neurotransmitters (see *neurotransmitter*).

Bibliography

[CYR 12] CYRULNIK B., *Un merveilleux malheur*, Odile Jacob, Paris, 2012.

[MCC 43] McCULLOCH W.S., PITTS W., "A logical calculus of the idea immanent in nervous activity", *Bulletin of Mathematical Biophysics*, vol. 5, pp. 115–133, 1943.

[MER 16] MERZENICH M., "Growing evidence of brain plasticity", *TED Conferences*, LLC, 2016.

[PRO 08] PROCHIANTZ A., *Géométrie du vivant*, Odile Jacob, Paris, 2008.

[PRO 14] PROCHIANTZ A., "On ne se baigne jamais deux fois dans le même sapiens", *Les Ernest Conferences "15 minutes pour changer notre vision du monde"*, Ecole Normale Supérieure, OHNK and Universcience, 2014.

Works published in English about neurofeedback and biofeedback

[BUD 08] BUDZYNSKI T., KOGAN-BUDZYNSKI H., EVANS J.R. *et al.*, *Introduction to Quantitative EEG and Neurofeedback – Advanced Theory and Applications*, Academic Press, Cambridge, 2008.

[COB 10] COBEN R., EVANS J.R., *Neurofeedback and Neuromodulation – Techniques and Applications*, Academic Press, Cambridge, 2010.

[COL 13] COLLURA T.F., *Technical Foundations of Neurofeedback*, Routledge, Abingdon-on-Thames, 2013.

[DEM 16] DEMOS J.N., *Getting Started with Neurofeedback*, Norton & Company, New York, 2016.

[EVA 07] EVANS J.R., *Handbook of Neurofeedback – Dynamics and Clinical Applications*, CRC Press, Boca Raton, 2007.

[HAM 11] HAMMOND C., GUNKELMAN J., *The Art of Artifacting*, ISNR Research Foundation, Miami, 2011.

[KHA 13] KHAZAN I.Z., *The Clinical Handbook of Biofeedback*, Wiley-Blackwell, Hoboken, 2013.

[ROB 08] ROBBINS J., *A Symphony in the Brain*, Grove Press, New York, 2008.

[SOU 11] SOUTAR R., LONGO R., *Doing Neurofeedback*, ISNR Research Foundation, Miami, 2011.

[STR 11] STRACK B.W., LINDEN M.K., WILSON V.S., *Biofeedback & Neurofeedback Applications in Sport Psychology*, Association for Applied Psychophysiology and Biofeedback, Wheat Ridge, 2011.

[SWI 10] SWINGLE P.G., *Biofeedback for the Brain*, Rutgers University Press, New Brunswick, 2010.

[SWI 15] SWINGLE P.G., *Basic Neurotherapy: The Clinician's Guide*, Springer International Publishing AG, Basel, 2015.

[THO 09] THOMPSON L., THOMPSON M., *The Neurofeedback Book*, Association for Applied Psycho–physiology and Biofeedback, Wheat Ridge, 2009.

[YUC 08] YUCHA C., MONTGOMERY D., *Evidence-Based Practice in Biofeedback and Neurofeedback,* Association for Applied Psychophysiology and Biofeedback, Wheat Ridge, 2008.

Works published in German about neurofeedback and biofeedback

[HAU 16] HAUS K.M., HELD C., KOWALSKI A. *et al.*, *Praxisbuch Biofeedback und Neurofeedback*, Springer-Verlag, Berlin, Heidelberg, 2016.

Works published in French about neurofeedback

[FOU 11] FOURNIER C., BOHN P., *Le neurofeedback dynamique – Quand notre cerveau apprend à mieux se réguler*, Dangles Editions, Escalquens, 2011.

Some studies about neurofeedback published by PubMed and concerning subjects addressed in this book

COMMENT.– All of these studies were conducted with first-generation neurofeedback systems. Studies are currently underway with *dynamic neurofeedback*. It is certain that their results will bring about better acceptance of this tool, which is currently criticized because of the opacity surrounding how the system works – an opacity that I have tried to dispel in this book.

[ARN 13] ARNS M., "The role of sleep in ADHD: possibilities for prevention of ADHD?", *Tijdschrift voor psychiatrie*, PubMed: 24166337, 2013.

[CAN 13] CANTERBERRY M., HANLON C.A., HARTWELL K.J. *et al.*, "Sustained reduction of nicotine craving with real-time neurofeedback: exploring the role of severity of dependence", *Nicotine & Tobacco Research*, PubMed: 23935182, 2013.

[COR 10] CORTOOS A., DE VALCK E., ARNS M. *et al.*, "An exploratory study on the effects of tele-neurofeedback and tele-biofeedback on objective and subjective sleep in patients with primary insomnia", *Research Unit Biological Psychology*, PubMed: 19826944, 2010.

[ENR 14] ENRIQUEZ-GEPPERT S., HUSTER R.J., SCHARFENORT R. *et al.*, "Modulation of frontal-midline theta by neurofeedback", *Biological Psychology*, PubMed: 23499994, 2014.

[GLO 13] GLOMBIEWSKI J.A., BERNARDY K., HÄUSER W., "Efficacy of EMG- and EEG-Biofeedback in Fibromyalgia Syndrome: A Meta-Analysis and a Systematic Review of Randomized Controlled Trials", *Evidence-Based Complementary and Alternative Medicine*, PubMed: 24082911, 2013.

[GRU 14] GRUZELIER J.H., FOKS M., STEFFERT T. *et al.*, "Beneficial outcome from EEG-neurofeedback on creative music performance, attention and well-being in school children", *Biological Psychology*, PubMed: 23623825, 2014.

[KOP 13] KOPŘIVOVÁ J., CONGEDO M., RASZKA M. *et al.*, "Prediction of treatment response and the effect of independent component neurofeedback in obsessive-compulsive disorder: a randomized, sham-controlled, double-blind study", *Neuropsychobiology*, PMID: 23635906, 2013.

[KOT 01] KOTCHOUBEY B., STREHL U., UHLMANN C. *et al.*, "Modification of slow cortical potentials in patients with refractory epilepsy: a controlled outcome study", *Epilepsia*, PubMed: 11442161, 2001.

[KUB 13] KUBIK A., BIEDROŃ A., "Neurofeedback therapy in patients with acute and chronic pain syndromes – Literature review and own experience", *Przeglad Lekarski*, PubMed: 24167944, 2013.

[LIN 12] LINDEN D.E., HABES I., JOHNSTON S.J. *et al.*, "Real-time self-regulation of emotion networks in patients with depression", *PLoS One*, PubMed: 22675513, 2012.

[NAN 13] NAN W., WAN F., LOU C.I. *et al.*, "Peripheral visual performance enhancement by neurofeedback training", *Applied Psychophysiology and Biofeedback*, PubMed: 24101183, 2013.

[OST 10] OSTERHAGEN L., BRETELER M., VAN LUIJTELAAR G., "Does arousal interfere with operant conditioning of spike-wave discharges in genetic epileptic rats?", *Epilepsy Research*, PubMed: 20388587, 2010.

[STE 14] STEINER N.J., FRENETTE E.C., RENE K.M. *et al.*, "In-School Neurofeedback Training for ADHD: Sustained Improvements From a Randomized Control Trial", *Pediatrics*, PubMed: 24534402, 2014.

[THO 02] THORNTON K.E., "The improvement/rehabilitation of auditory memory functioning with EEG biofeedback", *NeuroRehabilitation*, PubMed: 12016349, 2002.

[THO 13] THOMAS K.P., VINOD A.P., GUAN C., "Design of an online EEG based neurofeedback game for enhancing attention and memory", *Medicine and Biology Society*, PubMed: 24109716, 2013.

[WIN 01] WING K., "Effect of neurofeedback on motor recovery of a patient with brain injury: a case study and its implications for stroke rehabilitation", *Topics in Stroke Rehabilitation*, PubMed: 14523737, 2001.

[ZOT 14] ZOTEV V., PHILLIPS R., YUAN H. *et al.*, "Self-regulation of human brain activity using simultaneous real-time fMRI and EEG neurofeedback", *NeuroImage*, PubMed: 23668969, 2014.

Index

Printed in the United States
By Bookmasters